Jefferson County Library
620 Cedar Ave.
Port Hadlock, WA 98339
(360) 385-6544
www.jclibrary.info

STEAM

STEAM

*The Untold Story
of America's
First Great Invention*

62387
SUTCLIF

ANDREA SUTCLIFFE

Cover illustration of John Fitch's 1787 steamboat in Philadelphia is from a painting by John Franklin Reigart, from the Library of Congress, Prints and Photographs Division (LC-USZ62–1362). The artist was in error regarding both the date and the boat's name.

Left inset image: *Orukter Amphibilos,* Oliver Evans's amphibious steam vehicle, Philadelphia, 1805

Right inset image: Robert Fulton's first plan for a side paddlewheel boat, England, 1793

STEAM

First published 2004 by PALGRAVE MACMILLAN™
175 Fifth Avenue, New York, N.Y. 10010 and
Houndmills, Basingstoke, Hampshire, England RG21 6XS.
Companies and representatives throughout the world.
PALGRAVE MACMILLAN is the global academic imprint of the Palgrave Macmillan division of St. Martin's Press, LLC and of Palgrave Macmillan Ltd. Macmillan® is a registered trademark in the United States, United Kingdom and other countries. Palgrave is a registered trademark in the European Union and other countries.

ISBN 1–4039–6261–8 hardback

Library of Congress Cataloging-in-Publication Data
Sutcliffe, Andrea J.
    Steam : the untold story of America's first great invention / Andrea J. Sutcliffe.
       p.   cm.
    Includes bibliographical references and index.
    ISBN 1-4039-6261-8
    1. Fitch, John, 1743–1798. 2. Rumsey, James, 1743?–1792. 3. Marine engineers—United States—Biography. 4. Steam-navigation—United States—History—18th century. 5. Inventors—United States—Biography. I. Title.

VM140.F5S88   2004
623.87'22'0973—dc22

                                                                 2004040000

A catalogue record for this book is available from the British Library.

Design by Letra Libre, Inc.

First edition: July 2004
10 9 8 7 6 5 4 3 2 1

Printed in the United States of America.

For Ed

CONTENTS

ACKNOWLEDGMENTS

Many thanks to the helpful librarians in Philadelphia, Washington, D.C., and Richmond who helped me in my research, including Roy E. Goodman and Valerie-Anne Lutz at the American Philosophical Society; Phil Lapansky and Cornelia King at the Library Company of Philadelphia; Leonard C. Bruno at the Manuscript Division of the Library of Congress; and the staff of the Manuscript Reading Room, Library of Congress. Thanks also to librarians at the Historical Society of Pennsylvania, the West Virginia Historical Society, the Ohio Historical Society, the Virginia Historical Society, the University of Arkansas Library, the James Madison University Library, the Alderman Library at the University of Virginia, and the Library of Virginia. And to those who built the Library of Congress's American Memory web site, which makes it so incredibly easy to search and access the papers of George Washington and Thomas Jefferson, not to mention the many other useful documents about early America, a thousand thanks.

Thanks as well to Paula Johnson at the Transportation Division of the National Museum of American History, Smithsonian Institution, for providing access to material not available elsewhere, and to Jim Roan at the NMAH library. Fred Jaggi of the New England Wireless and Steam Museum in East Greenwich, Rhode Island, provided me with hard-to-find information about the early steam engine that operated near Cranston.

Many thanks to Nicholas Blanton, president of the Rumseian Society in Shepherdstown, West Virginia, for sharing the society's files on James Rumsey and his own extensive knowledge of Rumsey and his steamboat. Jeanne Mozier of Berkeley Springs, West Virginia, provided me with details of her research on Rumsey, which she used to create the informative exhibit on the inventor in her town's museum.

I would also like to thank friends and colleagues who read and commented on the manuscript: Jerry Fee, Patricia Slifka, Cicely Wynne, Tom Downing, E. Dean Vaughan, and Hugo Miller. Thanks as well to my agent, Ed Knappman, to editors Debbie Gershenowitz and Brendan O'Malley at Palgrave Macmillan, and to my mother, Nesta L. Johnson, for their enthusiastic support of this book. Thanks always to Paul Fargis. Finally, loving thanks to my husband, Ed, for everything.

INTRODUCTION

In the summer of 1787, two events that would change the course of human history converged in Philadelphia. The first, of course, was the writing of the U.S. Constitution. The second was the running of the world's first fairly reliable machine-powered vessel. To call it a steamboat would be correct, but it didn't look anything like the Mississippi riverboats that came later. Alongside the graceful sloops and schooners on the Delaware, it stood out, an ugly duckling among swans. A dozen crank-driven oars, mounted six to a side on a large wooden rack, creaked and groaned as they slowly pushed the boat along. The inventor of this odd contrivance was a tattered genius named John Fitch, who had started out wanting to build a steam-powered car.

Steam power was an idea whose time was at hand, the motive force the new nation needed to help secure its economic independence. After the Revolution, George Washington's chief concern was how to unify a country divided by a hundred-mile-wide mountain range. He was certain that trade was the key to binding the two halves, but the roads of the day were hardly more than trails, and carting goods across the Appalachians was difficult, slow, dangerous, and expensive. Across the mountains, the Mississippi and Ohio Rivers and their tributaries formed a network of natural superhighways. But without boats that could move upstream as easily as down, those rivers were one-way only.

In September 1784, less than a year after resigning his military commission, Washington crossed the mountains to examine road, river, and canal possibilities for connecting the Potomac River with points west. Five days into his trip, he met a Virginia millwright named James Rumsey, who showed him a model of a boat that could move upstream. Excited by the boat's potential, Washington publicly supported Rumsey's plan. A year later, he would reject John Fitch's plea of support for his steamboat idea. It may have mattered that Rumsey was a well-dressed, well-mannered southern gentleman, while Fitch was a straight-talking New Englander in threadbare clothes.

Washington warned Rumsey about Fitch, and the steamboat wars began. These two men spent the next several years fighting each other, their investors, politicians, and the world in general. With nothing to lose but their pride, they refused to work together, and they each refused to quit.

The scientific leaders of the day—Benjamin Franklin and David Rittenhouse in particular—didn't believe steamboats would ever work well enough to be practical. Franklin, in fact, would have been dismayed to learn that he inadvertently hindered steamboat development. In a paper he published on his return from France in 1785, just as the steamboat inventors were getting started, Franklin theorized that paddlewheels were an inefficient way to propel boats. He proposed using air- or water-jet propulsion instead. Because Rumsey and Fitch were desperate for Franklin's approval, neither man ever gave paddlewheels a try. Rumsey went for jet propulsion, while Fitch used paddles that dipped in and out of the water like the oars of a rowboat. Paddlewheels were never as efficient as the screw propellers that came much later, but given the weak engines of the day, they beat jet propulsion hands down.

In the spring of 1786, Fitch came to Philadelphia determined to build his steamboat. His first challenge was to devise a Watt-type steam engine without the benefit of ever having seen one. Although the English manufacturing firm of Boulton & Watt had recently begun selling steam engines, a new British law banned their export. Not a single Watt engine existed in America. Undaunted, Fitch formed a company of investors, hired a clockmaker as his lead engineer, and coaxed blacksmiths to make precision parts. In the new republic's heady first years, the belief that anything was possible and that individuals could make a difference (and perhaps a fortune) encouraged men like Fitch. "What cannot you do," he told himself at the beginning of his quest, "if you will get yourself about it."

Several months after Fitch's 1787 success on the Delaware, Rumsey came to Philadelphia to attack Fitch head on, fuming that Fitch had stolen his idea. Not long after, Franklin and several other influential Philadelphians formed a company to send Rumsey to England, where he could apply for a patent for his boat design and simply buy a steam engine. Just days after he arrived, Rumsey met with Matthew Boulton and James Watt. They were so impressed with his plans that they offered him a partnership. But after several rounds, their negotiations broke down and Rumsey returned to London to work alone. He spent the next few years dodging creditors while trying to build a water-jet-propelled steamboat on the Thames.

By 1790 Fitch had redesigned his steamboat and built an engine reliable enough to run regular passenger service between Philadelphia and Trenton. But even at eight miles an hour—twice the speed of Fulton's steamboats years later—he couldn't compete with stagecoaches. Most of his investors bailed out for good when Thomas Jefferson, as head of the first patent board, awarded Fitch and Rumsey federal patents dated the same day. Jefferson's unwillingness to choose one man over the other helped ruin both inventors. The government act that was meant to stimulate technological progress—and in fact was pushed into law to settle the steamboat wars—instead stifled it for many years.

Fitch and Rumsey persevered for a while longer, mocked by the public and dogged by bad luck. Rumsey died the day before his first scheduled steamboat trial in London. Fitch traveled to France for one last attempt but arrived to find himself in the middle of the Reign of Terror. The steamboat pioneers who followed—Oliver Evans, John Stevens, Samuel Morey, Nicholas Roosevelt, and a few others—faced similar hardships. In those early days of steam technology, most citizens viewed inventors as self-indulgent crackpots, not pioneers of progress.

In 1806, Robert Fulton, after a string of failures in other endeavors, returned to America after living twenty years in Europe and made the steamboat a success. He was a shrewd and ambitious businessman who had money and advantages the early inventors could only dream of. Fulton found the perfect partner in Robert R. Livingston, whose political pull gained them a twenty-year navigation monopoly on the Hudson River—geographically and commercially a perfect place to run a steamboat. Unlike his predecessors in steamboat building, Fulton by then was able to hire European immigrants who had mechanical expertise. Ignoring (or perhaps unaware of)

Franklin's warning, he used paddlewheels. By studying the failures and successes of the previous inventors, he knew what would work and what would not. Probably most important, a few years earlier he had craftily added a clause to his British military contract that allowed him to purchase and export a Boulton & Watt steam engine.

Although Fulton later received two U.S. patents, he never claimed to be the inventor of the steamboat. In fact, the head of the U.S. patent office in the early 1800s—a thorn in Fulton's side for years—often declared that Fulton's patents would be indefensible in court. But Fulton knew that navigation monopolies were better than patents, and he spent the last years of his life fighting to hold on to them. It would take a ruling of the U.S. Supreme Court nearly a decade after his death to strike down his Hudson River monopoly as unconstitutional. After that, the steamboat business boomed in the east and on the Mississippi. By the mid-1800s, some six thousand steamboats were running on the Mississippi.

The effect of steamboats on the nation's growth and economy was tremendous. Before they came along, it took four to six weeks to float a cargo-laden flatboat or barge downriver from Pittsburgh to New Orleans. It took another four months for the crew to walk back to their home port and start the process all over again. The high costs of labor, added to the product loss and spoilage that often occurred over such long periods, raised the price of goods to the point where they were nearly unaffordable. Trade languished. When the first Mississippi steamboat paddled up the river from New Orleans to Pittsburgh in 1815, it did so in twenty-five days, an enormous leap forward.

Steam engines transformed America. By the 1830s and for a century to come, they were powering not only boats and ships but railroad locomotives, mills, factories—and even cars. Productivity soared and prices dropped. The nation, united by trade made possible by easy transport, grew and prospered, as Washington had hoped. But it didn't happen overnight. What follows is the true story of the persevering and forgotten men who struggled to lay the groundwork in the nation's early days. It begins with the need for a boat that could push itself upstream and an idea for a steam-powered carriage.

ONE

STREAM-BOATS AND STEAMBOATS

*The natural mightiness of America expands the mind, and it partakes of
the greatness it contemplates.*

—*Thomas Paine*

Late on a drizzly September afternoon in 1784, George Washington
guided his horse down a rocky trail into the mineral springs town of
Bath. After slogging through thirty miles of muddy Virginia backcountry that day, the recently retired general was looking forward to an
evening under roof in one of his favorite places.[1]

A small crowd of people, the last to leave for the summer season, greeted
America's first hero as he entered the tree-lined town square; they had been
eagerly awaiting his arrival. Washington had not been in Bath, sometimes
called the warm springs of Berkeley, since 1769. On that occasion, he and
Martha had brought her daughter, Patsy, to take the waters in hopes of cur-
ing some unspecified "complaint."[2] The town had grown considerably since
then. A new tavern and boardinghouse, "at the sign of the Liberty Pole and
Flag," had recently opened on the square, and it was here that Washington

spent the next three nights.[3] This time his stay was purely business, part of a month-long trip west that he had been planning all summer.[4]

Among the townspeople welcoming the general was the part-owner of the tavern, a forty-one-year-old millwright and builder named James Rumsey. In a meeting described by an early Washington historian as "fortunate and undoubtedly prearranged," Rumsey at some point that evening spoke alone with Washington, and they agreed to meet the next day behind the tavern.[5]

At the appointed time, Rumsey, carrying what looked like a toy wooden boat, found Washington and led him through the woods to a fast-moving spot on Warm Springs Run. He knelt down and gently set the model in the water, its bow facing upstream. Washington looked on, first skeptically and then in amazement. The little boat—a convoluted arrangement of poles connected to a front-mounted waterwheel—slowly began to push itself forward against the current. At that moment Washington must have thought that fate had blessed his visit here. That evening he recorded the demonstration in his journal:

> ... [I] was showed the model of a boat constructed by the ingenious Mr. Rumsey, for ascending rapid currents by mechanism; the principles of this were not only shown, and fully explained to me, but to my very great satisfaction, exhibited in practice in private under the injunction of secrecy, until he saw the effect of an application he was about to make to the assembly of the state, for a reward.
>
> The model, and its operation upon the water, which had been made to run pretty swift, not only convinced me of what I before thought next to, if not quite impracticable, but that it might be to the greatest possible utility in inland navigation; and in rapid currents; that are shallow. And what adds vastly to the value of the discovery, is the simplicity of its works; as they may be made by a common boat builder or carpenter, and kept in order as easy as a plow. ... [6]

Rumsey's model looked something like a catamaran, except that a waterwheel was mounted above the bows of two connected boats. The flow of the downstream current turned the waterwheel, which was attached to a device on which setting poles were loosely mounted. The poles caught the river bottom and pushed the boat forward. This approach to working boats against the stream was purely mechanical; no engine was used.

Rumsey's idea—not original with him, according so some historians[7]—drew on the age-old practice of poling, in which men push long wooden poles into the riverbed to move boats upstream and guide them downstream. Poling a vessel against the current for long distances is possible but never fast or easy. In those days, most cargo boats and barges were sent downstream as one-way, one-use vessels that were broken up and used for building materials at their destinations. The crews would then walk many miles back home and start the process over again. Inefficient as this method was, it beat the alternative. Horse-drawn wagons made slow progress over the miserable roads of the day, and goods shipped overland could cost ten times as much as those shipped over water.

Rumsey was delighted to see Washington's favorable and even excited response. Seizing the moment, Rumsey asked the general for a statement in writing that he could use to help attract investors and seek funding from the state of Virginia. Washington was happy to grant the favor. That same day, Washington hired Rumsey to build a house, a kitchen, and stables on two lots he owned in Bath. He had little time to lose. His deed required that structures of a certain size be erected on each lot by November 1785, just over a year away.[8]

Washington had left Mount Vernon on September 1, accompanied by his close friend, Dr. James Craik. Leaving Bath, they were joined by his nephew, Bushrod Washington, and Dr. Craik's son, William.[9] They brought with them a tent, cooking supplies, fishing gear, six horses, and three servants. The purpose of this nearly seven hundred–mile trip, Washington told friends, was to visit his numerous landholdings across the Alleghenies. The journey would take him to into western Maryland and Pennsylvania, as far west as present-day Pittsburgh, and then back through parts of today's West Virginia. He had a long to-do list: visit overseers, evict squatters, and lease or sell some of his properties. Eight years of war, all of it away from Mount Vernon, had left his business affairs in shambles. Since resigning his army commission the previous December, Washington had spent many frustrating hours trying to get his records in order.[10]

In July, Washington had placed an ad in a Pennsylvania newspaper announcing an auction in Bath on September 7, at which he hoped to lease a large tract of farmland along the Potomac River twelve miles from town.[11] By his own tally, Washington owned nearly fifty thousand acres of land in Virginia, Kentucky, Ohio, and Pennsylvania. He had come by some of it years earlier, as payment for his service as a young officer in the French and Indian War. The rest was obtained by hiring agents to buy up thousands of acres from his fellow soldiers, who preferred ready cash to owning property in some remote and dangerous wilderness.[12]

Washington had a second and more visionary reason for this trip west: " . . . to obtain information of the nearest and best communication between the Eastern and Western waters. . . ."[13] His friends had long complained that he could spend entire evenings boring them with his thoughts on inland navigation—by which he meant a transportation network of rivers, canals, and roads. The Appalachian mountain range was the stumbling block. More than a hundred miles wide and running from Georgia to Maine, it formed a "castle wall . . . and the 'doors' of this wall were few and far between."[14]

He knew the area well. He began his working life in northern Virginia as a teenage surveyor for Thomas, the sixth Lord Fairfax, in the late 1740s; Lord Fairfax held title to more than five million acres of land in the Virginia colony. In 1753, Washington first crossed the Alleghenies as an emissary for the British army, and he later served in the Shenandoah Valley as colonel of the Virginia regiment during the French and Indian War. In 1759, at the age of twenty-seven, he resigned his military commission after being elected as the Frederick County delegate to the Virginia House of Burgesses. Even as a young man, he was impressed by the vast potential of the lands across the mountains. At one point he introduced a plan to the legislature that would improve the Potomac River for shipping; it died for lack of interest.

Fifteen years later, the idea had not left him. In 1774, with the help of his friend and future governor of Maryland, Thomas Johnson, he managed to get bills passed in Virginia and Maryland to form a company to make the Potomac navigable for a 150-mile stretch. The Revolution came along and put a halt to the project.[15] In 1783, while waiting at his military headquarters in Newburgh, New York, for official word that the treaty ending the Revolutionary War had been signed, Washington still had inland navigation on his mind. He made a trip to western New York to take a look at the geography of the area where, half a century later, the Erie Canal would be built. He wrote to a friend, "I . . . could not but be struck with the immense diffusion and importance of [inland navigation]; and with the goodness of that Providence which has dealt her favors to us with so profuse a hand. Would to God we may have wisdom enough to improve them. I shall not rest contented 'till I have explored the Western Country, and traversed those lines (or great part of them) which have given bounds to a New Empire."[16]

Washington wasn't alone in these thoughts. Earlier that year, Thomas Jefferson had written him a letter encouraging him to revive his dream of finding a way to connect the western reaches of the Potomac with the tributaries of the Ohio—and for Virginia's sake, to do it before the New Yorkers managed to join the Hudson River to the Great Lakes.[17] At every stop on this trip west, Washington questioned men who had explored the region, asking them for the smallest of details. His journal is filled with meticulous observations on distances, river conditions, and terrain. While in Bath, he noted that "from Colonel Bruce whom I found at this place, I was informed that he had traveled from the North Branch of the Potomack to the waters of the Yaughiogany, and Monongahela."[18]

Rumsey was, like many men of his time, a jack-of-all-trades, largely self-taught. He had moved from Maryland to Virginia about two years prior to his fortunate meeting with Washington. He was born in March 1743 at Bohemia Manor in Cecil County, on the eastern shore of the Chesapeake Bay. His father was a farmer, the third generation of Welsh Rumseys in Maryland. As a young man, Rumsey left the family farm and ran a tavern near the head of the Bohemia River that had belonged to his great-grandfather. At some point he moved across the Chesapeake Bay to Baltimore, where he was probably a millwright. A friend claimed he served in the Revolution, but no records survive to prove it.[19]

In 1782, Rumsey purchased land on Sleepy Creek a few miles from Bath, a town popular since the 1750s for its warm springs, which were said to offer "cures" for various ailments. (The town is better known today as Berkeley Springs, West Virginia, about a two-hour drive west of Washington, D.C. Its bathhouses, now part of a state park, are still open to the public.) Two years after the Revolution, Bath was experiencing a building boom. During his first few years there, Rumsey built and operated a sawmill on his Sleepy Creek property. For others he built gristmills, an iron mill, and several buildings in and around the town.[20]

By all accounts, Rumsey was personally charming and quite good-looking. A portrait of him made a few years later in London by the American artist Benjamin West shows a man with chiseled lips and large dreamy eyes. With powdered hair and ruffled shirt, he looks more like a poet than a millwright. He married twice. It seems likely his first wife died, because he mentions having a daughter, Susanna, when he married Mary Morrow of Shepherdstown, Virginia. In those days, Shepherdstown was a thriving community on the Potomac River, about thirty miles east of Bath. He and Mary soon had two children of their own. Rumsey apparently was drawn to the area by siblings who had moved there from Maryland. His sisters had married Shepherdstown men—one to his wife's brother, Charles Morrow, and the other to Joseph Barnes. Both brothers-in-law would become Rumsey's partners in boat-building.[21]

As soon as Washington left Bath, Rumsey went to work to gather support for his boat idea. He placed an ad in a Richmond newspaper, the *Virginia Argus,* that reproduced the general's glowing endorsement.

I have seen the model of Mr. Rumsey's boats, constructed to work against the stream; examined the powers upon which it acts; been eye-witness to an actual experiment in running water of some rapidity, and give it as my opinion (although I had little faith before) that he has discovered the art of working boats by mechanism and small manual assistance against rapid currents; that the discovery is of vast importance and may be of the greatest usefulness in our inland navigation, and if it succeeds (of which I have no doubt) that the value of it is greatly enhanced by the simplicity of the works which, when seen and explained, may be executed by the most common mechanic.

Given under my hand at the town of Bath, County of Berkeley, in the State of Virginia, this 7th day of September, 1784.

George Washington[22]

Although the records are fuzzy in this regard, it appears Rumsey had been tinkering with boat ideas for a year or two before he met Washington. On July 4, 1783, Rumsey's business partner, James McMechen, had asked the Continental Congress for a grant of land on the Ohio River should their invention—the mechanical poleboat—succeed.[23] But the members of Congress were "very apprehensive" that such a boat would ever work and turned him down. Not wanting to discourage inventive ideas completely, they urged him to try again in the future. A year later, and three months before Washington's visit to Bath, Rumsey and McMechen asked the Virginia Assembly for financial support. But the best the cash-strapped legislature could do was to offer "a sum adequate to the importance of this discovery" should the invention prove practical and successful.

Around the time of Washington's visit—and Washington may have been the one to tell him—Rumsey learned that states, instead of giving inventors money or land grants to support their efforts, would sometimes award term monopolies, in which inventors were allowed exclusive use of their products in the state for a set time, usually ten to fourteen years. They were the closest things to patents at the time. Rumsey set out that fall and winter of 1784 to seek state monopolies for his mechanical poleboat—which he called a *stream-boat,* causing historians and proofreaders endless confusion for years to come. Within a few months, Virginia, Maryland, and Pennsylvania granted him monopolies for what presumably was the stream-boat. The wording of his applications was vague, though, perhaps deliberately so. He was unsuccessful in New Jersey, New York, North Carolina, South Carolina, and Georgia.[24]

While Rumsey was on the road, Washington returned to Mount Vernon more convinced than ever of the need for a western trade route. He wrote in his diary that Virginia must "open a wide door, and make a smooth way for the produce of that country to pass to our markets before the trade may get into another channel." In those days before the thirteen states were bound by a strong federal government, citizens tended to think of their home state first, in competition with the others. Here, Washington still had the possibility of a New York canal on his mind. But he also talked of the importance of easy transportation to the nation as a whole, the western boundary of which extended to the east bank of the Mississippi River. With France and Spain holding the lands on the other side, he was concerned that as Americans continued to move west, they would be more likely to trade with those two countries. He worried that the United States might then split in two. The only force that could hold the country together, he believed, was trade between its eastern and western lands.

In a summary Washington wrote of his trip, he saw Rumsey's boat as a timely blessing: " . . . Rumsey's discovery of working boats against stream by mechanical powers principally, may not only be considered as a fortunate invention for these States in general but as one of those circumstances which have combined to render the present epocha favorable above all for securing . . . a large portion of the produce of the western settlements and of the fur and peltry of the lakes, also—the importance of which alone, if there were no political considerations in the way, is immense."[25]

A week after his return, Washington sent a long letter on the issue of inland navigation to Virginia Governor Benjamin Harrison urging that the state move quickly; he mentions Rumsey's boat idea as being a "very fortunate invention." Harrison quickly wrote back to tell him he had presented Washington's letter to the legislature. By November, a young delegate named James Madison had introduced several bills dealing with canal and river navigation.[26]

By chance, Rumsey and Washington were both in Richmond that November, lobbying the assembly for their respective causes. Once Rumsey learned the general was in town, he wasted no time in tracking him down. Probably lying in wait outside some meeting room, Rumsey finally spotted Washington and pulled him aside. In hushed tones, he told the general that he was experimenting with steam as a source of power for a new kind of boat, quite different from the model he had demonstrated in Bath. Washington appeared to listen politely. But a year later, when pressed to recall the conversation, Washington said he hadn't been paying close attention and couldn't quite recall Rumsey's words.[27] This private conversation would come back to haunt the secretive Rumsey.

Later that month, Madison wrote to Jefferson describing what had happened that month in Richmond. He noted that when Rumsey had asked the assembly the previous spring for financial help to develop a boat that would move ten miles an hour upstream, its members rejected him with scornful laughter ("The apparent extravagance of his pretensions brought a ridicule upon them, and nothing was done.") But, Madison continued, as soon as the delegates learned of Washington's endorsement, "it opened the ears of the Assembly to a second memorial," which resulted in an act that gave Rumsey a ten-year monopoly in the state.[28]

Washington's influence was golden elsewhere as well. Thanks to his personal appeals, Virginia and Maryland quickly passed legislation in January 1785 authorizing the Potomac Navigation Company.[29] Its charter was to make the river navigable west of what is now Washington, D.C., to a point west of Cumberland, Maryland, largely by removing rocks and boulders and by building short canals and connecting roads where river travel was impossible. Within weeks, subscription books were set up in cities throughout Maryland and Virginia so that private citizens could buy shares to raise money for the venture. James Rumsey, with one share, was one of the first subscribers.[30]

The winter of 1784–85 passed with Rumsey tending to his businesses in Bath and, by his account, experimenting with various forms of propulsion and tinkering with steam engines. A close friend and resident at his boardinghouse, one Nicholas Orrick, stated a few years later that he had witnessed one of Rumsey's "curious experiments" that winter. It consisted of a long, hollow wooden tube, rigged with weights and pulleys. As Rumsey poured water down the tube, "it drew up a certain weight . . . I asked Mr. Rumsey what he meant or intended by this experiment, to which he made answer, that by that principle he would make the boat go."[31] It appears from this description that Rumsey had run across the theories of the celebrated European scientist Daniel Bernoulli.

Half a century earlier, Bernoulli had published a paper, "Hydrodynamica," which among other things set forth the idea of propelling a boat by forcing a jet of water through a tube from the front of the boat and out the back, using the force of the water to propel the vessel forward—a simple form of jet propulsion. In 1784, Benjamin Franklin, coincidentally, was thinking thoughts similar to Rumsey's. About a year later, in December 1785, he would present a paper in Philadelphia that, among many other things, endorsed jet propulsion as the best way to propel a boat. During his years in Europe, Franklin had come to agree with scientists there who believed that waterwheels and, by extension, paddlewheels, were inefficient. The idea of paddlewheels on boats was hardly new.[32] A drawing from ancient Rome depicts a paddlewheel boat turned by oxen on board, and the Chinese apparently sailed human-powered paddlewheel boats in the twelfth century. Even Leonardo da Vinci, in 1482, sketched designs for a human-powered paddlewheel boat.[33]

Rumsey turned to another ancient idea, the force of steam, as the way to power the pump that would propel his jet boat. Steam was first used as a power source in ancient Greece but was never put to any serious use. In the 1500s and 1600s steam was occasionally mentioned in scientific treatises, but it wasn't put to practical use until 1681, when the Frenchman Denis Papin invented the pressure cooker. About ten years later, Papin figured out how to power pistons by steam.[34] The first steam-powered machine, a water pump with an unfortunate tendency to explode, was invented in 1698 by Thomas Savery in England.[35] Around 1712, another Englishman, Thomas Newcomen, invented a different kind of steam engine, which used the pressure of the atmosphere to push a piston that in turn drove a pumping mechanism. Over the next few decades Newcomen engines, several stories tall, began pumping water out of mines all over Europe. Although Newcomen's engine was bulky and inefficient, it was less likely to blow up because of its use of valves and low pressure.

In the years before Rumsey began to think about steam power, three Newcomen engines existed in America. The first was imported from England and installed by an English engineer, Josiah Hornblower, at the Schuyler copper mine in Belleville, New Jersey. This engine operated from

1755 to 1760, when it was damaged by a fire. Repairs were made, and it was put back into operation until another fire destroyed it in 1768.[36] The second Newcomen engine in America was built by an Irish immigrant named Christopher Colles in New York City to pump a public water supply for the city.[37] It was completed and working, "albeit rather imperfectly," in April 1776. But when the British captured New York, the project was abandoned.[38]

Colles had been experimenting with steam for a while before that. Two years earlier he had presented his ideas on steam engines to the American Philosophical Society and had tried to build a steam engine that would raise water for a Philadelphia distillery. He could never get it to work reliably. The third engine in America was built sometime after 1780 by Joseph Brown and put to work pumping water from an iron mine near Cranston, Rhode Island. It operated into the early nineteenth century.[39]

Coincidentally, 1784—the year of Washington's visit to Bath and Rumsey's first experiments with steam—was also the year that the Scotsman James Watt and his partner Matthew Boulton began manufacturing their double-acting steam engine in Birmingham, England. This version was the most efficient Watt engine to date. Yet only a few Americans, mostly those who had traveled in Europe or knew someone who had, were even aware of it. In fact, the saga of steamboat development in America might have progressed at a quicker pace if only some enterprising boatbuilder been able to order a steam engine from the company of Boulton & Watt, which is exactly what Robert Fulton would do two decades later. But in 1785, Britain, not about to share new technology with her rebellious former colony, enacted laws that prohibited the export of technology and the emigration of skilled workers to America. In those days, anyone interested in building steam engines in America was working in a vacuum.

That was Rumsey's situation in 1785. Although he had probably heard of Watt's work, the only steam engines he knew anything about were the Savery or Newcomen variety. A Newcomen atmospheric steam engine is a relatively simple machine. A large boiler heats water and produces steam, which flows into a cylinder. As the steam expands, it pushes a piston—which is connected to one end of a large overhead rocking crossbeam—to the top of the cylinder. Then a valve opens and sprays cold water into the cylinder, which condenses the steam and creates a partial vacuum. The pressure inside the cylinder decreases, and the atmospheric pressure outside the cylinder forces the piston back down. A heavy pump piston is connected to the other end of the rocking crossbeam, so that when the cylinder piston falls, the pump piston is pulled to its highest point, raising water through a pipe from the mine below. The weight of the pump piston causes it to fall back down, the action of which pulls the cylinder piston back up, and the process is repeated. This back-and-forth motion is termed reciprocal, and steam engines of this type are called reciprocating. Newcomen engines were highly inefficient, though, because the cylinder had to be alternately heated and cooled.

From skimpy descriptions and sketches, Rumsey set out to build, using crude materials and local tradesmen, a steam engine of his own design. He had at least one technical reference: a copy of Desaguliers's *A Course of Experimental Philosophy,* published in 1744, which included drawings and descriptions of a Savery steam pump.[40] (Desagulier made improvements to the Savery engine in the early 1700s.) Rumsey's original plan is unknown, but by 1787 he had come up with the following design, as revealed by his application for a Pennsylvania patent that year. A large boiler passed steam into the topmost of two vertical cylinders, each of which contained a piston joined by a single rod. As steam filled the upper cylinder, it caused both pistons to rise, drawing river water through valve-controlled openings in the hull. When the upper piston reached the top of its stroke, the steam in the upper cylinder was expelled through a tube to a condenser, creating a vacuum. The piston in the upper cylinder would then drop, also pushing down the piston in the lower cylinder, which in turn forced the river water through a tube in the hull and out the stern. No paddles were used; the force of the ejected water pushed the boat forward.[41]

Rumsey's clever design addressed two of the biggest problems of Newcomen engines: their size and their efficiency. Newcomen engines were huge, meant for use on solid ground. Even then, the engine caused "a tremendous noise and trembling of the large building in which it is erected," according to a 1787 account of the Cranston steam engine mentioned earlier. So much power would be needed to move the monstrous engine that little would be left to move the boat—assuming it didn't sink first. Making the engine smaller and more efficient was crucial. Rumsey's combination engine and pump solved the size problem; his pipe boiler improved efficiency.

The early steamboat inventors faced one more design problem: If they wanted to use paddles to propel the boat, they had to devise a way to convert the reciprocal motion of the steam engine's pistons into rotary motion. Rumsey sidestepped this problem by using his engine to power a water pump, the original purpose of a Newcomen engine.

Evidence exists for only two attempts to build steamboats before 1785, both in Europe. In 1736, Jonathan Hulls received an English patent for a Newcomen-driven steamboat. Although he published a pamphlet describing the boat, not much is known about it other than his apparent attempt to build and test it. A drawing in his pamphlet shows a boat with a paddlewheel mounted at the stern, which the inventor placed there for a scientific reason: because ducks and geese "pushed their web feet behind them."[42] In France in 1776, the Marquis de Jouffroy d'Abbans, probably also using a Newcomen engine, unsuccessfully tested his first steamboat. In 1783 he tried again, this time with limited success with a larger boat on the Saône River near Lyon. His *Pyroscaphe* used two paddlewheels mounted on an endless chain. It ran for fifteen minutes before the vibrations of the engine tore the boat apart.[43]

In December 1784, Rumsey's business partner, James McMechen, returned to New York to plead once more for encouragement from the Continental Congress.[44] This time, the delegates were more agreeable. Like the Virginia Assembly members, they were swayed by Washington's praise for Rumsey. They offered the men a land grant of 30,000 acres if their boat was a success, which was defined as being able to carry ten tons of cargo with a crew of three men and to travel up the Ohio River an average of fifty miles a day for six consecutive days.[45]

Before approving the grant, though, Hugh Williamson, a member of the committee formed to look into the matter, wrote a skeptical letter to Washington asking him if Rumsey's boat worked as claimed. Washington answered that it did: " . . . if a model, or a thing in miniature can justly represent a greater object in its operation, there is no doubt of the utility of the invention."[46] Good words from the general continued to benefit Rumsey, and Congress passed a resolution supporting his boat the following May.[47] Around that time, though, Rumsey's run of good luck began to fizzle. In April, a fire at his mill near Bath destroyed the lumber intended for four of his construction projects, including Washington's large house. Rumsey gave Washington the bad news, along with a bill for work done so far on the "small houses" on his lots, in a letter in late June.[48]

At the time, Washington was busy drumming up subscribers for the Potomac Navigation Company. He had reluctantly agreed to serve as president of the company, and his first job was to find a qualified engineer to lead the effort. Despite a heavy ad campaign announcing the job—he had even sent Rumsey a few handbills to post in Bath—by July no suitable candidates had turned up. Desperate to keep the project moving and impressed with Rumsey's creative abilities, Washington hit on the idea of recruiting the inventor. In early July, he wrote to him, "As I have imbibed a very favorable opinion of your mechanical abilities, and have had no reason to distrust your fitness in other respects; I took the liberty of mentioning your name to the Directors." He encouraged Rumsey to gather some letters of recommendation and apply for the job.[49] Flattered, Rumsey solicited a few references, including one from Revolutionary War General Horatio Gates, who was then living near Shepherdstown. But certainly all Rumsey needed was Washington's endorsement.

At a meeting of the Potomac Navigation Company in Georgetown in mid-July, the board of directors hired Rumsey as superintendent and Richardson Stewart of Baltimore as his assistant. Rumsey was told he could hire another assistant and as many as a hundred laborers. To make up for lost time, the plan was to put two teams to work, one near Harpers Ferry, the other between Seneca Falls and Great Falls. Rumsey was elated to have such a prestigious job and a steady income, but he wasn't ready to abandon his boat experiments. To keep things moving, he hired his brother-in-law Joseph Barnes to complete the full-size boat he had been building.

Rumsey's headaches began just four days into the job. In a letter to William Hartshorne, the company treasurer, Rumsey was already pleading for money to pay for supplies. Personnel problems took up most of his time. Although pay and benefits were good, it was hard to find men willing to do the backbreaking labor of blasting and removing rocks, clearing river bottoms, and digging canals. Worse, the men knew this. On the whole, according to Rumsey's letters, they were insolent, did sloppy work, and were drunk and disorderly a good part of the time. By August, barely a month after work began, Rumsey wanted to fire half of them.

In frustration, he sent a letter to Washington and the board of directors describing his problems. The board thought they might obtain a more dedicated group of workers by purchasing sixty indentured servants from European ships landing in Baltimore. By October, the board also agreed to hire one hundred "good and able Negroes" who would receive the same pay and benefits as everyone else. The company was eager to get as much work done as possible before the river froze.

But the new workers did not solve Rumsey's problems. He complained again to Hartshorne in late September. When pay was late, he explained, " . . . they get a little drunk, I am cursed and abused about their money in such a manner that contrary to my wish, I am obliged to turn abuser." People living in the vicinity began complaining to both Rumsey and the sheriff, saying they had been terrorized by the rough characters working on the river. Several of Rumsey's men were arrested and jailed, while many others simply walked away. To his great dismay, Rumsey was spending far more time on labor and supply problems than on engineering.[50]

In his few spare moments, he dabbled with boat and engine ideas. He claimed to have made his first attempt at a steam boiler—really just a large iron pot and lid, with most of the hands-on work carried out by Barnes. Barnes finished their boat near Bath in the fall of 1785 and brought it downriver to Shepherdstown, where he set out to get parts made for the boiler. That December, he brought the boat and the machinery down the Potomac to Shenandoah Falls, near Harpers Ferry. When Rumsey had time, he helped Barnes and McMechen work on the boiler. But within a few weeks the Potomac iced up, forcing them to pull the boat in for the winter without testing the engine.[51]

In February 1786, problems with the pot boiler led Rumsey to experiment with what he would call his pipe boiler. At some unknown date, he instructed the Antietam Ironworks to forge iron in the shape of tubes. The tubes were threaded and joined with threaded collars. The man who helped weld them would later state that the tubes sat in his shop for at least six months, until Barnes showed up and directed him to bend them "nearly in the shape of a worm of a still, with this difference, that the rounds were placed so close, as nearly to touch each other."[52]

The idea—a brilliant one—was to fill the pipes with water and set them inside a furnace, where their large surface area, much greater than that of a pot, would be exposed to the fire. Rumsey notes, in an article he published two years later in the *Columbian Magazine,* that the pipe boiler's advantages were its compact size, its fuel efficiency, and its ability to force water "much higher than any other kind." In its first tests, though, the pipe boiler let steam escape from the seams, a flaw that Rumsey blamed on poor workmanship, not bad design.[53]

Depressed at his lack of progress on the job and with his steam engine experiments, Rumsey soon faced another worry. In February 1786 he received an ominous letter from Washington strongly urging him to go public with his boat ideas. "I will inform you further than many people in guessing your plan have come very near the mark, and that one who had something of a similar nature to offer to the public, wanted a certificate from me that it was different from yours. I told him I was not at liberty to declare what your plan was, so I did not think it was proper to say what it was not."[54]

Rumsey's problems had just begun.

TWO

A Ridiculous Idea

In the availability of men willing to persevere with a possibly "ridiculous"
idea, America had an advantage.

—*Frank D. Prager*

Washington's disturbing letter was sparked by an incident that had occurred almost four months earlier. For some reason, perhaps a reluctance to become involved further, he delayed telling Rumsey.

The encounter took place, Washington wrote in his journal, on an unseasonably warm Friday evening in early November 1785.[1] A rough-looking stranger arrived at Mount Vernon around dusk and asked to speak with the general. He certainly would have mentioned that the general's old friend, Governor Thomas Johnson, had sent him. Washington, ever the gentleman, agreed to see the man and led him into the parlor. At two inches over six feet tall, John Fitch of Bucks County, Pennsylvania, was about the same height as the general, with a straight-backed posture to match. At forty-one, he was twelve years younger, a fact that his thin and unhealthy appearance disguised. His clothes were dusty and torn, and his straight black hair touched his shoulders. But his dark eyes were intense and intelligent. Washington motioned Fitch to sit down.[2]

Not a man for small talk, Fitch immediately began to describe a revolutionary device of his own creation—a steam-powered boat. He rattled off mechanical details: cylinders, boilers, pistons, paddles, speeds. He made claims for his boat's enormous importance to the nation's expansion west, where long, wide rivers like the Ohio and Mississippi could be traveled upstream as easily as down. A few minutes into his spiel, Fitch noticed that Washington seemed to show "some agitations of the mind that was not expressed." Sensing the problem, Fitch "requested of him to know if it was the same plan as Rumsey's, [but] he informed me that could not give Rumsey's plan by negatives." Not knowing what to say, Washington got up to leave. He told Fitch he needed some time to think things over and would return in a while.[3]

"Perhaps an hour after, he came into the room where I was, as if by design, and informed me that my plan was not the same as Rumsey presented to him at Bath, but that some time after that at Richmond he had mentioned something of the sort to him, but he was so engaged in company that he did not attend to it." Fitch switched the subject to the reason for his visit. He asked the general for a letter of introduction to the Virginia Assembly. Washington refused, offering no explanation. Fitch left Mount Vernon angry and upset, passing on the general's offer to stay the night. He was relieved to hear that Rumsey's boat was not yet powered by steam, but he was wounded by Washington's refusal to help.[4]

Why Washington rejected Fitch outright is puzzling. He may have been put off by Fitch's appearance and rough manner. It's possible that Fitch may have aroused unhappy memories of an unstable and incompetent inventor of Washington's acquaintance named John MacPherson, who lobbied Congress unrelentingly during the early years of the Revolution to support his untested military defense schemes.[5] Perhaps—although there is no evidence for this—he was bothered by a vague memory of Fitch as a profiteering supplier of beer and rum to his troops at Valley Forge during the Revolution. Less likely, he may have known that Fitch had been a land speculator in Kentucky and the Ohio Valley.[6] Washington by then held such landjobbers in contempt; two years earlier, he referred to them as "a parcel of banditti."[7] Most probably, though, Washington was leery of lending his name to any more dubious boat enterprises. As far as he could tell, Rumsey had made little or no progress on his boat projects that year and indeed had been full of excuses as to why, after more than a year, his house at Bath still wasn't finished.[8]

John Fitch was born in Windsor, Connecticut, on the Hartford line, on February 1, 1744.[9] Like Rumsey, Fitch grew up on a hardscrabble farm. He was a great-grandson of one of five Fitch brothers who came to America in the 1600s from Essex, England. Much more is known about Fitch's early life than Rumsey's because Fitch left behind a detailed autobiography.[10]

His childhood hardships were probably typical of those harsh times, but John was a sensitive child, suffering hurts early and often. His mother died when he was four, leaving his father to raise John and his five siblings. Four decades later, Fitch remembered the event: " . . . believ[ing] her death to be the greatest loss of which I ever met with. . . . [E]arly in the fall I was deprived of her and altho I did not consider my loss natural, affection carried my grief to a very great excess for a child of my age." His father remarried a few years later and John was raised by a benevolent stepmother.

When John was six, he "suffered a most extraordinary circumstance" that "seemed to forebode the future." He had come home from school to find his older brothers and sister working in the barnyard. He entered the dark house, where his younger sister, Chloe, was eager to show him some new toy. She lit a candle to look for it, and in doing so set fire to two tall bundles of flax. John ran to pick up the sheaves, one at a time, to carry them to the safety of the hearth. In the process, his hair was singed and his hands were burned. Chloe ran outside and yelled to an older brother that John was setting the house on fire. About that time, John ran out of the house, crying in pain. The older boy fell on him and beat him. When their father came home late that evening, John, bruised and burned, pleaded for understanding but was ignored.

Years later, Fitch excused his father's harsh behavior by explaining that "he was educated a rigid Christian and was a bigot and one of the most strenuous of the sect of Presbyterians." He claimed his father carried his beliefs to such excess that as a child he didn't dare pick an apple off a tree on the Sabbath. His father's main duty, as Fitch saw it, was to make sure his children learned the Bible. That done, "if I could earn him 2d a day it ought not to be lost."

When John was eight or nine, his father pulled him out of school and put him to work on the farm. The boy loved school: "I was nearly crazy with learning. . . . This piece of injustice I can never forgive him." But his father didn't discourage book learning, so John studied any book he could find, one of the first being his father's copy of *Hodder's Arithmetick*. When he was about eleven, he heard of a book that "would give me information of the whole world which was Salmons Geography."[11] The elder Fitch refused to buy the book, but the boy persevered, finally convincing his father to let him plant potatoes on a patch of unused field. The following spring, he sold his crop for ten shillings and asked a Windsor merchant to order the book for him. When it arrived months later, he devoured it. "At that time no question could be asked of me of any nation but I could tell their numbers, religion, what part of the globe, their latitude and longitude." A few years later, John's father softened a bit and agreed to let him go to school for a few weeks each winter. He also bought John some tools to learn surveying.

When he was nearly eighteen, Fitch recounts how he had been left out of a community project to build a new meeting house in Windsor: " . . . it was a very serious affront offered to me . . . that I never really got over." He knew it was time to leave Windsor. On the day when most of the townspeople had gone to watch the meeting house steeple being raised, John borrowed a horse and rode down to the Delaware River to look into a life at sea. He asked a captain to let him have a trial run, unpaid, on a ship to New York. The captain told him he'd have to cover his own costs. To John's surprise, his father approved and even gave him several dollars for the trip. "Such unaccountable generosity in him raised the highest sense of gratitude in me to so kind a father and I set off with plenty of riches to try the seas." The ship, however, sailed to Rhode Island, not New York, and passed through a violent storm on the way. Five weeks of bad weather convinced Fitch that a life on a ship was not for him.

Not long after returning to work on his father's farm, Fitch ran into a man named Benjamin Cheany, who was looking for an apprentice in his clock making business. He could hardly believe his good fortune. With some apprehension, Fitch agreed to a deal in which he would pay for his training by working at odd jobs for the family. These tasks soon took up all of his time and included humiliating "woman's work" such as washing dishes and milking cows. Worse, Cheany refused to teach him anything about clock making.

This so-called apprenticeship—apparently not unusual for the time—continued for two and a half years, during which Fitch complained endlessly. Eventually Cheany, weary of the young man's constant carping, agreed to let Fitch serve out the remaining year of his service with his brother Timothy, also a clockmaker. But all Fitch learned from Timothy was brassmaking. It seems the Cheany brothers had no desire to train a potential competitor. In his three years as a clockmaker's apprentice, Fitch never once was allowed to see a clock or watch being made. On his twenty-first birthday, he was free to leave. "I sat out for home and cried the whole distance and doubt not but nearly as much water came from my eyes as what I drank. I acknowledge that this is an unmanly passion but cannot to this day avoid such effeminacy."

For the next two years he lived with his father and earned a fairly good living making small brass items. With his savings, he and two partners formed a business to build a potash works; he used potash in his brassmaking business and figured he could easily sell any excess. He moved to a village about twenty-five miles from home to set up the works. While living in a boardinghouse there, he became acquainted with his landlady's sister, Lucy Roberts, a woman he described as older than himself, "in no ways ugly" but something of an old maid. After six months, they married, on December 29, 1767. Fitch's delicately phrased explanation for this decision appears to suggest a little premarital fooling around: "I cannot say that I was ever passionately fond of the woman but for the sake of some promises made in haste and for some favours granted in consequence of said promises I determined to marry her."

Ten months later their son, Shaler, was born. A little over a year later, Fitch had had enough of Lucy's nagging and told her he was leaving, unaware she was pregnant again. Fitch describes a heartbreaking scene in which Lucy followed him out of the house and down the road for quite a distance, pleading with him to stay.

Filled with guilt and misgivings, and with no plans for his future, Fitch wandered from town to town, cleaning clocks door-to-door to support himself. He was either supremely confident or totally desperate; his only experience with clock cleaning had been a one-time try, a fortunate success, on a neighbor's clock. After six months of life as a tramp, he reluctantly decided to go to Brunswick, New Jersey, and enlist in the British army. But even the army rejected him, probably because he "had the appearance of one being considerably advanced in the consumption."

By mid-May 1769, Fitch had settled in Trenton, where he eventually turned a series of small jobs making brass and silver buttons into a profitable silversmithing business with several employees. Fitch recalled that by the time the Revolutionary War began in 1775, he had managed to amass savings worth $4,000, a small fortune in those days. A committed patriot, Fitch joined the New Jersey militia as a second lieutenant. When his mechanical abilities became known, he was put to work as a gunsmith. For an upcoming battle in Amboy, Fitch was told to gather guns from local citizens. Working day and night, Fitch and a group of helpers were able to provide arms for about fifty soldiers within three days.

Fitch was sent to the battle as a lieutenant in command of a company. When his troops arrived in Maidenhead, Fitch was denied a promotion that he felt was rightfully his. "The resentment and mortification which I felt on the occasion I had never experienced before. I then took my gun and knapsack on my back and marched back to Trenton alone and wished to have my judgment convince me that Great Britain was right. But finally concluded not to resent my wrongs on my country for the rashness of a few inconsiderate boys."

In Trenton, Fitch continued repairing and refitting arms. As the war effort grew, he and his men worked from dawn to midnight, seven days a week. When the elders of Fitch's Methodist church learned that he was

working on Sundays, they expelled him. This event reinforced Fitch's growing disenchantment with established religion.

When the British arrived in Trenton, Fitch's militia unit was called up to
fight. He became entangled in a political struggle and was denied command
of a company, even though the superior officer had assigned him to the post.
In a snit, he deserted and fled to Bucks County to hide out for a while. He
began to write a letter pleading his case to his commanding general, but before he could mail it word arrived that Washington's army had defeated the
Hessians at Trenton. With this good news, Fitch dropped the idea, figuring
he would rejoin his unit later if they needed him.

Upon returning to his house in Trenton, he found that the British had destroyed most of his furniture and belongings. He packed up what was left and
moved to Warminster, then a village of about four hundred people, in Bucks
County, Pennsylvania, hoping for a fresh start. To make money, he began
supplying beer and tobacco to the British army in Philadelphia in the fall of
1777, which turned out to be a highly profitable venture. The following winter and spring he provided beer, rum, and other items to Washington's troops
at Valley Forge. When the army left Valley Forge in 1778, Fitch returned to
silversmithing but this time barely made a living at it. Inflation was rampant,
and within a year Fitch's savings had shrunk in value to $40.

Searching for a new career, Fitch began to listen more closely to stories
about the fortunes that could be made by surveying uncharted land in Kentucky, then part of Virginia. Under a state act passed in 1779, individuals
could purchase land warrants, at a price of £40 per hundred acres, that authorized them to survey a specified quantity of land in certain areas. After
the survey was made, it would be recorded with the state, which would later
legally grant the land. Landjobbers turned this process into a business, with
the goal of subdividing the acreage and selling lots at huge profits to settlers.
With recommendations (and financial backing) from a few well-to-do Penn-
sylvanians, Fitch obtained a Virginia commission as a deputy surveyor in
Kentucky from none other than James Madison. Fitch left Pennsylvania in
the spring of 1780, and after escaping from a band of hostile Indians on the
trip down, Fitch surveyed "the choicest lands and in the heart of the country," about sixteen hundred acres worth, and returned the following spring

to record the land in Richmond. Thoughts of an "immense fortune" filled his head.

By March 1782, Fitch was ready to return to Kentucky for a second surveying expedition. To raise money for the trip, he bought several barrels of flour, which he planned to take by barge from Pittsburgh down the Ohio and the Mississippi Rivers and resell in New Orleans. On March 22, four days into the trip, Fitch and the men he was traveling had a run-in with Indians—in the attack, a bullet whizzed past Fitch's face—and were taken prisoner. After a few weeks of forced marching, Fitch and his companions were detained in an Indian town, where they feared for their lives. Somehow they had learned that they were "the first prisoners taken after Williamson's Massacre . . . [and] for my own part expected nothing but retaliation," according to Fitch. Presaging a decade of bad timing for Fitch, just two weeks before he was captured, a Pennsylvania man named John Williamson had led nearly a hundred militiamen in an attack on a group of Christianized Delaware Indians in Gnadenhutten, Ohio, in revenge for Indian attacks in Pennsylvania. He and his men brutally murdered ninety-six Delawares, thirty-nine of whom were children. This atrocity sparked a wave of Indian attacks on white settlers in the Ohio Valley over the next several years.

Fitch and his companions were threatened but not killed. They were eventually marched to Detroit, where the Indians turned them over to the British. From there they were transferred to a military prison on an island near Montreal. With some basic tools provided by his keepers, Fitch busied himself by planting a garden, making brass and silver buttons, and repairing watches. He was allowed to sell his wares, and "in about four months I got to be as rich as Roberson Cruso." Shortly after, his captors announced that they were ready to release him. But for once in his life he was fairly content. He begged to be allowed to stay, but the British refused, laughing at his request.

They put Fitch on a ship bound for Boston. It was late fall, not the best time of year to travel, especially for someone inclined to seasickness. Plans changed along the way and the ship sailed to New York instead, arriving on Christmas Day of 1782, after a stormy ten weeks. Ill, hungry, and lice-infested, and still held as a prisoner on the ship, he wrote to two friends, fellow Methodists who lived in New York, and asked them to bring him a few supplies. They ignored his pleas, which "gave me such a disgust to Christianity." Sick and exhausted, he was eventually freed and made his way home to Warminster.

The following year, with a peace treaty between the United States and Great Britain in the works, Fitch predicted that Congress would soon be looking to sell the lands north of the Ohio River and east of the Mississippi (the area known as the Northwest Territory, made up of today's Ohio, Indiana, Illinois, Michigan, Wisconsin, and part of Minnesota). He saw a chance to survey the choicest lands and so be among the first to apply for land warrants the minute Congress opened up the area. He formed a landjobbing company of several prominent men, including Dr. John Ewing, provost of the University of Pennsylvania, and William C. Houston, who was then a member of the Continental Congress. He may have been introduced to these men by another member of his company, his friend and pastor at the Neshaminy Presbyterian Church, Nathaniel Irwin.

During the summer and fall of 1783, Fitch and his team crossed the Appalachians and surveyed thousands of acres in the Ohio Valley. That winter he returned to Warminster to plan a return trip the following spring, which turned out to be his most productive. "On the whole that trip we made out locations for about 250 thousand acres of most valuable lands when I returned well satisfied, being morally sure I should one day or other become a man of fortune."

While he was surveying that spring and summer, Fitch heard the news that Congress had approved the Territorial Ordinance of 1784. This act, which never became law, was written by Thomas Jefferson and described how the western lands should be governed and subdivided. He learned that official surveyors would soon be appointed for this new land, so when he returned to Warminster that fall he applied to Congress for one of the positions.[12] By now, Fitch was desperate for a steady income. He felt sure of success because of his familiarity with the territory and the impressive references he provided.

That winter of 1784–85 in Warminster, "having nothing to do and for my own amusement [I] sat to and made a draft of the country [the Northwest

Territory] from Hutchins' and Murrow's maps with the additions of my own knowledge." It struck him that this more accurate map of the new U.S. lands could be turned into a source of income. "I got a sheet of copper and hammered it polished it and engraved it and then made a press and printed it."[13]

In early spring he got word that the Pennsylvania surveyor job went to someone else. A few months later, he received even worse news: Congress had passed the Land Ordinance of 1785, which called for the land in the Northwest Territory to be sold by Congress in lots to the public, not granted to land speculators like Fitch. "By said resolve from an immense fortune [I] was reduced to nothing at one blow." Now his only hope was a half-baked idea for a steamboat that had hit him that April.

John Fitch's steamboat idea was barely eight months old when he approached Washington in the fall of 1785. In the spring of that year, Fitch, a seeming failure at the age of forty-one, took a walk that changed his life. On his way home from the Neshaminy Presbyterian Church one Sunday, his knee, afflicted with rheumatism since his days in Indian captivity, seized up in pain. He paused and cursed, knowing he would walk the remaining mile in agony. He describes in his autobiography what happened next:

> . . . And in Street Road a gentleman passed me in a chair with a noble horse. A thought struck me that it would be a noble thing if I could have such a carriage without the expense of keeping a horse. A query then rose immediately in my mind. Thus viz what cannot you do if you will get yourself about it. . . . I soon thought that there might be force procured by steam and set to and made a draft. And in about one week's time gave over the idea of carriages but thought it might answer for a boat and better for a first rate man of war.[14]

Fitch wasn't the first person to imagine a powered carriage, a dream of a machine that would one day evolve into the modern automobile. Unknown to Fitch, another Pennsylvania man, Oliver Evans, had been thinking about the idea since 1772; in 1786, he asked the Pennsylvania Assembly to award him a patent for a steam carriage. The delegates laughed him out of the hall, and he gave up on the idea for many years.[15] In England, James Watt wanted to put his steam engine to use on a carriage in the 1770s but thought better of it. The first real attempt was made in 1769, when a French military engineer named Nicholas-Joseph Cugnot built a three-wheeled steam vehicle that featured a gigantic copper boiler mounted ahead of the front wheel. It achieved a speed, if you could call it that, of a little over two miles an hour. Unfortunately, it had to stop every ten minutes or so to build up steam, a process that took another twenty minutes. It was also quite difficult to brake and steer. One day Cugnot lost control and crashed his graceless heap into a stone wall, putting it out of its misery.

Fitch, luckily, knew none of this. He worked night and day on his idea for a few weeks, and when he thought he had a solution he walked back to the Neshaminy church to run it past Reverend Irwin.[16] Fitch admired Irwin more as a scholar than as a man of the cloth, once telling him that he "felt a secret pain that such an exalted genius should be confined to the pitiful business of Neshaminy congregation." He hoped Irwin, a Princeton graduate, could review his plans for a steamboat and tell him how to proceed.

Irwin welcomed Fitch into his office and took the papers his friend handed him. He carefully examined each drawing, probably half listening as Fitch rambled on about the possibilities. After a few minutes, Irwin got up and searched his bookshelf, finally pulling out a worn volume entitled *Philosophia Britannica*.[17] He found the page he was looking for and showed it to Fitch. On it was a drawing of a Newcomen steam engine, remarkably similar to Fitch's sketch.

Fitch was "amazingly chagrined," as he put it later, when he examined a drawing of a device that had been in use for decades. "Till then," Fitch recalled later, "I did not know there was such a thing in nature as a steam engine."[18] Although upset to find out his idea was not original, he was comforted by the thought that it was feasible. "It strengthened my opinion in the scheme, knowing that the machinery could not fail of propelling if I could gain the force, as my only doubts lay in gaining the force itself."

Fitch spent the summer figuring out how to connect an engine to a paddle device. He ran his ideas by every mechanic or educated man he could find in Bucks County. One day in August he walked the thirty or so miles down York Road to Philadelphia, then America's largest city and its intellectual and cultural capital. (A woman in Warminster who knew him remembered Fitch as being "a great walker": "he could always out-travel a horse. In walking he pitched forward, and went onward with a great swing." She recalled that he once walked forty miles in a day.[19]) His purpose was to call on a few of his influential landjobbing partners to ask for funding and letters of support.

Fitch's first stop was at the University of Pennsylvania, where he showed his plans to Dr. Ewing. Fitch explained that he hoped to persuade Congress to support his idea and perhaps even fund his efforts. Ewing was impressed and wrote a letter on Fitch's behalf to another land company partner and a member of the Continental Congress, William Houston. It read, in part, "Should such a machine be brought into common use in the inland navigation through the United States, it would be exceedingly advantageous in transporting the productions of America to market, and thereby greatly enhance the value of our back lands."[20]

Fitch took the letter to Houston, a Trenton lawyer. As it turned out, Houston was no longer a congressman, but the previous year he had served on the congressional committee that looked into giving James Rumsey support for his mechanical poleboat. Houston was probably the first to tell Fitch about Rumsey, and he certainly would have mentioned that the Congress had promised Rumsey thousands of acres western lands should his mechanical boat prove successful. Houston offered to write Fitch a letter of introduction to an influential member of Congress from New Jersey, Lambert Cadwallader. With Houston's help, Fitch hoped to convince Congress to give him the same deal it had offered to Rumsey a few months earlier.

A few days later Fitch dropped in on another acquaintance, Dr. Samuel Smith, provost of Princeton University. He walked out with another letter of introduction, this one to a North Carolina congressional delegate.

By the end of August, just five months after the idea of steam power hit him, Fitch arrived in New York City, the seat of the Continental Congress, armed with plans, testimonials, and a model of his boat. His petition began, "The subscriber begs to lay at the feet of Congress an attempt he has made to facilitate the internal navigation of the United States, adapted especially to the waters of the Mississippi."[21] With these opening words, Fitch unwittingly said the wrong thing. He was unaware that many members of Congress were more interested in protecting their business interests in the East than in navigating the Mississippi. Opening up that river, which was then controlled by Spain, would mean a loss of business to East Coast farmers and merchants. Nevertheless, a committee was formed to look at Fitch's petition.

Within a few days, the committee returned Fitch's papers, telling him his proposal wasn't worth submitting to the full Congress.[22] Fitch was furious, but he had a backup plan. He delivered a letter to the Spanish consul to New York, Don Diego de Gardoqui, in which he offered his steamboat idea to Spain. The ambassador immediately invited Fitch to visit him. Gardoqui told Fitch he was impressed with his plans, but that if Spanish government helped him, it would want exclusive rights to the boat. Fitch was taken aback. He may have been angry at Congress, but he wasn't ready to deprive his country of the benefits of his invention. He withdrew his offer and walked out. Dejected, he returned to Warminster and went back to work on his steamboat ideas, drawing inspiration from his desire to "revenge myself on the committee of Congress, and prove them to be but ignorant boys."[23]

In late September, Fitch returned to Philadelphia to formally present his papers, as well as a small model, to the American Philosophical Society. Founded by Benjamin Franklin to "promote useful knowledge" in 1743 (around the time Fitch and Rumsey were born), the society had become the preeminent gathering place for American men of science. Fitch, who had received so little schooling, needed assurance from its well-educated members that his plan was sound. He also wanted to make sure that his ideas became a matter of public record, should he ever need to prove that he was the first to have had them.[24]

He told the members on the evening of September 27 that his goal was "to gain their opinions, and find out if there was something had escaped my notice, which would render the scheme impracticable." The minutes of that meeting did not record the members' comments, but Fitch noted that no one brought up anything he had not already thought of. Franklin was not at the meeting that night. He had returned to Philadelphia just two weeks earlier after serving nearly nine years as American minister to France.

The papers and drawings he presented to the society have been lost, but the wooden model he showed to the members remains in its collection. It measures 23 inches long by 4¾ inches wide. On its right side is a device that demonstrates Fitch's mode of propulsion: an "endless chain passing over screws or rollers to which floats or blades were attached to answer as paddles."

While in Philadelphia, Fitch paid a personal visit to Franklin, who among his many other achievements was America's most accomplished scientist and inventor. By then he was seventy-nine, still involved in politics and other matters but slowed down by health problems. Fitch wrote that Franklin invited him in and "spoke very flatteringly of the scheme, and I doubted not his patronage of it, altho I could obtain nothing from him in writing." Coincidentally, Franklin told him, he had been working on his own ideas for boats, and he planned to deliver a paper on the topic to the society on December 2. Franklin explained that he would not support any particular plan until his own ideas had been set forth. It seems likely, though, that Franklin told Fitch about the ideas of the European scientist Bernoulli and his theories on various propulsion methods, including water-jet propulsion and screw propellers.

Not put off by Franklin's gentle rejection, a week later Fitch wrote him a letter dramatically restating his case:

> The subscriber most humbly begs leave to trouble you with something further on the subject of a steam boat. . . . And if his opinion carries him to excess, he doubts not but your excellency will make proper allowance. As it is a matter of the first magnitude not only to the United States, but to every maritime power in the world, as he is full in the belief that it will answer for sea voyages, as well as for inland navigation, in particular for packets where there should be a great number of passengers. . . . [N]othing would give me more secret pleasure than to make an essay under your patronage, and have your friendly assistance in introducing another useful art into the world.[25]

As soon as he returned to Warminster, he packed up to leave for Kentucky, where he intended to fight a claim on a large tract of land he had surveyed there four years earlier. On his way, he planned to stop over in Richmond to make sure his 1781 surveys had been properly recorded. On October 20, Fitch headed south, thoughts of steamboats filling his mind. By now he knew enough to realize that building a steam engine was "far beyond my abilities," so he continued to seek out anyone with a knowledge of science or mechanics. Passing through Lancaster, Pennsylvania, he called on William Henry, a gun maker and inventor (most notably of the screw auger).

Henry welcomed Fitch into his home, listened for a while, and dropped a bombshell. Henry explained that he had come up with the idea of steamboats in the mid-1770s but never did more than sketch out a few plans. He recalled discussing the idea with Thomas Paine, the American patriot and pamphleteer who dabbled in inventions, the scientist David Rittenhouse (who later referred to the steamboat and several other wild ideas as being "ridiculous"),[26] and the surveyor, astronomer, and mathematician Andrew Ellicott.[27] But his plans to build a working model were overtaken by the Revolution. Henry assured Fitch that the idea was a sound one and strongly encouraged him pursue it. He even promised Fitch that he wouldn't make claims for the idea himself since he had never gone public with it. Fitch was surprised at Henry's revelation but grateful for his offer of help.[28]

Back on the road, Fitch headed south to Fredericktown (today's Frederick, Maryland). Here he decided to work the political angle by dropping in on former Maryland Governor Thomas Johnson, George Washington's long-time ally in inland navigation schemes. Rumsey's name must have come up immediately. Johnson told Fitch that Rumsey was a customer at one of his iron foundries and that he knew all about his experiments with the poleboat. They discussed its limitations, including the fact that the boat required a certain amount of manual assistance and that it would not work in still or deep water.

Johnson praised Fitch's ideas and urged him to do two things: present his steamboat plans to the Virginia Assembly, which was then in session in Richmond, and seek Washington's endorsement. Johnson must have told Fitch that Washington had endorsed Rumsey's mechanical boat the previous year. Fitch "heartily concurred" with the first suggestion. But he was hesitant to impose on so great a man as Washington. "I had no expectations from [doing so], not doubting in my own mind but his excellency was heart sick of boat projects."[29]

In the end, Fitch decided to seek Washington's support because he did not want to offend Johnson. His only solace after that uncomfortable evening at Mount Vernon was that he had predicted the outcome. He later referred to the visit as "the most fatal thing to me in this scheme," taking deep offense at what he saw as unfair treatment from Washington. In his autobiography, written several years later when Washington was president, Fitch wrote, "I believe [Washington's] greatest failure is too great delicacy of his own honour, which we can hardly suppose can be carried to excess. The certificate which he gave to Mr. Rumsey respecting the pole boat, and [his] permitting him to publish it so freely as he did, was perhaps one of the most imprudent acts of his life, and undoubtedly touched him in the tenderest part. . . . How hard it is . . . to suggest that so amiable a character would injure an indigent citizen to save his own reputation."[30]

Leaving Mount Vernon, Fitch probably muttered curses at Washington all the way to Richmond, where he found a more encouraging reception. He described his plans to James Madison—"a person almost without deception, and the great statesman"—who agreed to present Fitch's petition to the Virginia Assembly. But the outcome was disappointing. The members offered praise and encouragement, but not in writing and with no offer of funding.

Persistent, Fitch took his case directly to the governor, Patrick Henry. "[He] appeared to be much pleased with the novelty and magnitude of the scheme, and wished to give it every encouragement which lay in his power." Fitch then proposed a method of funding that would spare the state's coffers (which were empty anyway) while giving him money to get started.

Governor Henry liked the idea: Fitch would sell subscriptions—in effect, advance sales—for the map he had made of the Northwest Territory to the assembly delegates, who in turn would sell subscriptions in their districts. In the days before publishing companies, people who wanted to sell printed items like maps or books would sign up buyers and get payment in advance, giving the author the money to cover printing and distribution costs. Henry signed a certificate, dated November 16, 1785, to be affixed to each subscription, that promised Fitch would pay £350 to the state when one thousand map subscriptions were sold.[31]

With this small success, Fitch canceled his trip to Kentucky and headed home to start printing maps. On his way back he stopped in Fredericktown to report his progress to Johnson. The former governor was pleased and bought ten subscriptions to Fitch's map. The two men again talked about Rumsey's plans. Fitch left feeling assured that Rumsey was not experimenting with steam. Rumsey would later claim, though, that he had built his first steam engine that fall. Fitch assumed that Johnson probably knew what Rumsey was doing all along, and he believed for years after that Johnson meant to deceive him. Johnson, though, saw nothing wrong in lending his support to both men and happily wrote a letter of introduction for Fitch to take to the then governor of Maryland, William Smallwood. He closed it with the words, "You will soon perceive that he is a man of real genius and modesty."[32]

Returning home to Warminster, Fitch learned that the Pennsylvania Assembly was in session and immediately went to Philadelphia to ask the delegates for help. Once more, he received encouragement but no certificate of support or offer of financial reward. While in town, he decided to try Franklin again, figuring that he could at least impress him with the news that Governor Henry was supporting his efforts.

He went to Franklin's home and asked—perhaps even begged—him for a letter of support. But Franklin offered moral support only and even refused to buy a subscription to Fitch's map. After listening to Fitch plead his case for a while, Franklin rose and went into another room. He called Fitch to come in. Fitch looked on as Franklin took five or six dollars from a desk drawer and offered the money to him. "The indignation which inflamed my blood could hardly be suppressed," Fitch wrote of the incident. "Yet I refused with all the modesty that I was master of, and informed him that I could not receive money unless it was monies that should be subscribed for, was I to do that I laid myself liable to censure of embezzling monies that was given for other purposes." Fitch later berated himself for his restraint and good manners. "I esteem that one of the most imprudent acts of my life, that I had not treated that insult with the indignity which he merited, and stomped the paltry ore under my feet."[33]

A month or so later, in early January 1786, he traveled to Annapolis to visit Governor Smallwood, handing him the glowing letter of introduction Johnson had written. He petitioned the Maryland Assembly for rights, and he also asked for funds to buy a Boulton & Watt steam engine from

England. It appears that Fitch and Andrew Ellicott, who was serving in the legislature at the time, had discussed the possibility of getting an export permit. This is the first mention in Fitch's papers that indicates he knew about Watt's engine—had Franklin told him?—but his timing was terrible. About that time, the news that Britain had enacted a law forbidding exports of technology to its former colony reached America.[34]

The Maryland delegates offered encouraging words but, like those in Virginia and elsewhere, turned down his request for financial aid. Before returning to Pennsylvania, Fitch dropped in on the Delaware Assembly, where he found only a few members in attendance. There, he simply talked about his plans and left.

THREE

Brother Saintmakers

A man's useful inventions subject him to insult, robbery, and abuse.

—Benjamin Franklin

While Fitch was busy rallying support for his steamboat idea that winter of 1785–86, problems continued to dog Rumsey in his job as superintendent of the Potomac Navigation Company. Two letters, written nine months apart to company officer William Hartshorne, reveal Rumsey's deteriorating mental state.

The first, in September 1785, describes a rock-blasting accident in which one of his best men had "both of his hands most horribly maimed by a charge of powder going off. . . . I hope the president and the directors will take his case into consideration and allow the poor fellow something for his lost time."[1] Nine months later, Rumsey had hardened. In his final letter to Hartshorne, on July 3, 1786, he wrote matter-of-factly, "We have been much imposed on in the last two weeks in the powder way (we had our blowers, one run off, the other blown up) we therefore was obliged to have two new hands put to blowing."[2]

After serving a few days short of one year, Rumsey had had enough. At a meeting of the company's board of directors, he announced his "disinclination to serve the Company any longer for the pay and emoluments which had been allowed him." The directors refused to increase his pay, so he quit. They promptly replaced him with Richardson Stewart, his assistant.

In August, still fuming that the board didn't value him enough to keep him on, Rumsey presented the directors with a list of ten charges against Stewart. He described Stewart's incompetence in mechanics; his dishonesty, insubordination, and cruelty to the workers; and his role in worsening the problems with the nearby residents. Rumsey believed his troubles had been made worse by a backstabbing assistant, and he wanted the company, especially Washington, to know it. The board promised to investigate his charges and report its findings at the next meeting.

When Rumsey attended the meeting that October, he must have been devastated to see words like "unfounded," "frivolous," "unsupported," and "not proved" throughout the report. It concluded by noting that only one of the ten charges had any merit at all. Worse, Washington's estimation of Rumsey appears to have slipped a few notches; he wrote in his diary that Rumsey's claims appeared "malignant, envious, and trifling." Two years later, however, Rumsey would be vindicated. The board's minutes noted that the members "removed Stewart for reasons relative to the interest of the Company," with "sundry charges of a serious nature brought against him."[3]

Rumsey, in the meantime, was able to devote most of his time to steamboat experiments. Even while he was supervising workers at Great Falls, he and Barnes had continued to tinker with the pipe boiler. But after numerous repairs, the primitive engine continued to fail, getting in only a few strokes before stopping dead in the water. In the spring of 1786, Rumsey gave up on it and went back to testing his pot boiler on a boat in the Potomac at Great Falls. It really was a pot—a large black kettle made by the local blacksmith, capped with a lid secured by rivets and solder. In a test, the boat moved forward for a few minutes, but soon steam began escaping, robbing the engine of the little power it had managed to work up. Nevertheless, Rumsey was encouraged that the boat moved at all.

That same spring, interestingly, a future steamboat inventor named Robert Fulton came to Bath to take the waters for a lung ailment. Just twenty years old at the time, Fulton was trying to support himself as a miniature-portrait painter in Philadelphia. There is no record that Fulton ever met Rumsey or his partners while he was in Virginia, but it seems likely that he would have heard stories in Bath about the inventor and his boat. According to Fulton's first biographer, Cadwallader Colden, Fulton befriended one or two influential people during his stay there who suggested that he continue his art studies in England. A year or so later, he left for London.[4]

By summer's end, Rumsey and Barnes, frustrated with leaky boilers, went back to experimenting with the mechanical poleboat. The first test, by Barnes on September 9 in Shepherdstown, drew a large crowd of curiosity-seekers. He let about nine or ten excited people on board, but their "shouting and running backward and forward" made an already unsteady craft even shakier, and just about everything that could go wrong did.[5] The poles were too light and slipped on the river bottom, which caused the wheel to come out of the water and cut power to the poles. All told, the boat moved forward about two hundred yards. Several days later Barnes and Rumsey tried again, but by then the water levels in the Potomac had dropped and the current was too weak for the wheel to generate much power. After a few minutes of embarrassingly slow progress, they decided to quit before a crowd gathered to witness yet another humiliating failure. Around this time the townspeople began referring to the inventor as "Crazy Rumsey."

In late September, Rumsey sent a progress report to Washington. Trying to remain optimistic, he noted that his mechanical poleboat "would make tolerable progress in all currents that is straight and clear of rocks, and moves three miles per hour, or upward, but will go but slow in currents under that velocity." In other words, a redesigned boat would work well in rivers that were fast running, not too shallow but not too deep, rock-free, and straight. This clearly was not a boat for most of the known navigable rivers in America. Ever positive, though, he ended the letter by inviting Washington to a demonstration at a place of his choosing.[6]

In December, Rumsey decided to try the pipe boiler again. He tested an improved version on a boat in Shepherdstown, and this time it worked a little too well. "The violence of the heat was so great, from the steam, that it melted the soft solder that a great part of the machine was put together with, and rendered it entirely useless, until repaired with hard solder; about this time, the ice drifting, carried off the boat which the machine was made for, and destroyed her in such a manner that the repairing her was equal to one half the expense of building a new one."[7]

While Rumsey sat out the winter, John Fitch was busy with plans to build a full-size steamboat. In February 1786, he set off for Philadelphia to find a lead engineer. His first choice was a man named Arthur Donaldson, who had invented a floating crane a decade earlier that was powered by horses and used to dredge river bottoms. It was called the "Hippopotamus." For this useful machine, Pennsylvania had recently awarded him a period of exclusive rights.[8]

Fitch explained his scheme to Donaldson and offered him a share of the profits in return for his help. Donaldson seemed much taken with the idea and gave Fitch the impression that he had never heard of a steamboat until their conversation. But Donaldson stalled; he asked Fitch for some time to think the offer over. The next time Fitch was in Philadelphia, Donaldson promised, he would have an answer for him.

While Fitch waited for Donaldson's decision, he went to Trenton and petitioned the New Jersey Assembly for exclusive navigation rights for his steamboat. During his years in Trenton as a silversmith before the war, Fitch knew many of the town's influential people. One was a well-to-do businessman named John Cox who promised to help Fitch financially if his New Jersey rights came through. Fitch thought he had finally hit the mark. To ensure success with the assembly, he drew up a certificate citing his ability to build the boat and asked fourteen respected New Jersey citizens to sign it. Leading the list were Cox and Cox's son-in-law, another wealthy businessman and sometime inventor named John Stevens.[9]

Fitch's efforts paid off. In late March, New Jersey granted him exclusive rights for fourteen years. But his high hopes evaporated when he returned to New Jersey several weeks later. Cox seemed to have forgotten his promise of support and handed Fitch a paltry ten dollars. Stevens, who had also promised Fitch money, gave him nothing.

This blow to Fitch's plans had been preceded by another betrayal earlier that month. Back home in Warminster, Fitch struck up a conversation with a woman from Philadelphia who was visiting his landlady. At some point Fitch told her about his invention. She responded with something along the lines of "What a coincidence—some people I know in Philadelphia were just talking about a man who was doing the same thing, and they thought it would be the greatest thing in the world!" She added that the man was talking about getting the state to award him exclusive rights for his invention. *What was the man's name?* Fitch asked. *I'm not sure I remember,* she replied. *Could it be Donaldson?* Fitch tried. *Yes,* she said, *that's it, that was his name.*[10]

Fitch wrote that this news "alarmed me considerable and determined me to set off for Philadelphia the next day." He headed straight for Donaldson's house and spoke with the man's wife. She confirmed that her husband did have plans to build a steamboat but said that his approach was different from Fitch's. Fitch was deeply disturbed. "I am ashamed to say that I did not close my eyes to sleep the ensuing night." It appears that some time that day, though, Fitch learned about someone who might be able to help him.

At dawn the next morning, Fitch was walking the frosty streets of Philadelphia in search of a stranger's boardinghouse. He found it and knocked on the front door "on or before 7 o'clock," as noted in the March 10 diary entry of John Hall, an English engineer who had recently moved to Philadelphia and the object of Fitch's visit.[11] Hall at that time was helping Thomas Paine, the patriotic pamphleteer of the 1770s, build models of new designs for wood and iron bridges. Both Hall and Paine considered themselves inventors; Paine had discussed his ideas for a steamboat, it may be recalled, with William Henry ten years earlier.

Hall wrote that night in his diary, "A brother saintmaker [Hall's term for inventors] came with a model of a machine to drive boats against the stream. Poor man had tears in his eyes." Fitch poured out his problems to Hall and to Paine, who also was staying at the boardinghouse. He begged them for advice and support. The two men tried to discourage Fitch in his plans to take his battle with Donaldson to the assembly, but Fitch refused to change his mind on that topic. Paine declined to help, saying he was too busy with his own work, but he did buy one of Fitch's maps and offered some ideas for simplifying his steamboat.

Hall did not record what, if any, suggestions he offered Fitch, nor did Fitch mention any offers of help from Hall. But Hall's knowledge of steam engines was most certainly what Fitch was after that morning. Hall, who was fifty years old at the time, had moved to Philadelphia from England the previous year. He has been described as "a close associate of the pioneers of steam power, Boulton and Watt, . . . [and] a free-thinking man of science."[12] His nephew and adopted son, John "Jack" Capnerhurst, who took the surname Hall after his adoption, had come to America two years earlier and was operating an iron foundry in New Jersey. This younger John Hall had worked in Boulton & Watt's steam engine workshop in Birmingham between 1778 and 1780. Watt eventually came to distrust him and urged Boulton to fire him, claiming that he was a "great blunderer" and was using their workshop to develop his own inventions.

By the end of 1782, Boulton had more serious concerns. He had learned that the younger Hall was spreading Boulton & Watt trade secrets.[13] Hall's diary, which along with his uncle's diary is in the collections of the Library Company of Philadelphia, contains technical data and drawings of steam engines. Was the elder Hall sharing his and his nephew's knowledge of Watt steam engines with Fitch?

The day after Fitch visited Hall and Paine, he returned to the Pennsylvania Assembly to petition for the exclusive right to use steam in navigation on state waters. He told the delegates that he feared another man was about to beat him to it, using a machine similar to his but with "some trifling alterations in the mechanism" that were intended to deprive him of his rights. The news of his plea traveled fast. The following day, Donaldson introduced his own petition to the assembly.

Fitch had learned that Donaldson planned to use steam to power a pump that would use air- or water-jet propulsion to propel a boat—the very plan suggested by Benjamin Franklin the previous December. To Fitch, this fact was good news. Not only could he claim priority of invention because he had presented his plans to the Continental Congress and the American Philosophical Society several months earlier, but he also could argue that Donaldson's idea was not his own but Franklin's. Also, he planned to argue that the legislators must consider the *means* of propulsion—whether water jet, oars, or paddlewheel—to be secondary to the key aspect of the invention: the use of steam as the source of navigational power.

Fitch vowed to fight Donaldson with every bit of ammunition he could lay his hands on. His first thought was to get a copy of Franklin's paper, "Maritime Observations," which included his ideas for machine-propelled boats.[14] The paper had been read at the December meeting of the American Philosophical Society but had not yet been published. He walked over to the offices of the American Philosophical Society and spoke to one of its officers, Francis Hopkinson, a respected Pennsylvania judge and a signer of the Declaration of Independence. Hopkinson told Fitch he did not know the whereabouts of the paper and sent Fitch to see the "secretaries of the society, thinking there I must get bewildered and lose sight of them," Fitch wrote. But one of those secretaries told Fitch that Franklin had ordered the paper delivered to Hopkinson. Fitch took this news to Franklin, who told him to tell Hopkinson to give the paper to Fitch.

When he returned to Hopkinson with Franklin's request, the man refused, saying he could not act on an oral order. Fitch went back to Franklin, explaining that Hopkinson wanted a written order. Franklin refused to give him one and instead offered to present his paper in person to the assembly when the time came. By now Fitch was fed up with the bureaucratic runaround he was being put through. "I was determined . . . to speak in a pretty positive manner, as I found nothing else would do." Some standard Fitch ranting and raving no doubt ensued; he later regretted his outburst. Franklin eventually gave in and agreed to let Fitch "extract the part [of the paper] which I wanted." Franklin wrote a note for Fitch to take to Hopkinson.[15] Once again Hopkinson refused to give Fitch anything, instead promising to bring the paper to the assembly. At his point, Fitch grew enraged and angrily accused Hopkinson of supporting Donaldson. With that, Hopkinson turned over the paper.

Fitch returned to the assembly with Franklin's plan and other documents to support his claim of priority. He urged the delegates to investigate the matter. The next day, Donaldson followed with documents of his own. The assembly agreed to form a committee to decide the case. Since the current session was about to adjourn, Fitch was told that its members would make a report in the next session.[16]

Feeling confident that the Donaldson threat would evaporate, Fitch set out once more to raise money and to try to find someone who could help him build an engine ("I never was vain enough to suppose myself equal to that task"). This time he approached John Nancarrow, who agreed to help. Nancarrow operated an iron foundry in Philadelphia, but a few years earlier in his native England he had worked on Newcomen-type steam engines. Nancarrow said he needed time to draw up plans, so while he waited, Fitch set out to form a company of shareholders. By mid-April, he had drawn up the necessary papers and issued forty shares of stock, keeping twenty for himself. In a week he sold nearly all the shares at a price of $20 a share.[17] In general, his supporters were not the richest men in Philadelphia, but rather shopkeepers, farmers, mechanics, and teachers. A few, including his former landjobbing partner Dr. John Ewing, were members of the American Philosophical Society.

In April 1786, Fitch packed up and moved from Warminster into a boardinghouse in the Northern Liberties section of Philadelphia, just north of today's historic area. With some forty thousand people, Philadelphia was by far the largest and most vibrant city in America, much larger than Boston or New York. As the country's leading port city, it had everything Fitch needed: a river to experiment on, a convenient source of labor and materials, and a population of potential investors. Ironically, around the time Fitch arrived, the man who would later get all the credit for inventing the steamboat, Robert Fulton, had left his job in a jeweler's shop at Second Street and Walnut and was working out of an art studio on Front Street at Pine, one street up from the waterfront. Whether he ever saw Fitch testing his steamboat is not known. But this was the second time, and it would not be the last, that Fulton crossed paths with the early steamboat inventors. Late that year or early the next (the exact date is not known), Fulton boarded a ship for England to pursue his career as an artist.

Fitch, flush with cash for the first time in years, was ready to get started but was still waiting to hear from Nancarrow. After Fitch sent a few reminders, Nancarrow replied with a set of drawings that disappointed Fitch greatly. "[It] was to work upon the old-fashioned plan of engines [Newcomen's], and to have a weight to raise the piston," Fitch described. "I was too dogmatical to be led into that plan, and had assurance enough to set up my opinion against the one who was esteemed the greatest engineer in America. In which I became positive and determined to do nothing unless it could be effected on a better plan."

He had two more possibilities. He decided to approach the only other men in America he knew of who had hands-on experience with steam engines: Josiah Hornblower and Christopher Colles. He asked his shareholders to approve his expenses for a trip to New York, where he hoped to convince Hornblower (who was then serving in Congress) or Colles to join him.

By chance, a few days before his planned departure for New York he met a Philadelphia clockmaker named Henry Voigt. Describing him as "a handsome man and a man of good address and familiar friendly and sociable to all and a truly honest man in his trade," Fitch saw in Voigt the "first marks of genius" and thought that he possessed technical abilities that could be applied to steam-engine building. Fitch canceled his trip and instead sent identical letters to Colles and Hornblower (starting with the words "Being an entire stranger to steam engines . . .") explaining his plan and asking for their comments.[18] After his experience with Nancarrow, he wasn't expecting much; after all, neither man had ever made significant improvements to the Newcomen engine. He heard back from Hornblower, who offered encouragement and a few suggestions.[19]

Voigt, on the other hand, was so eager to help that he insisted on working for free. Fitch complained that he had to talk Voigt into accepting a share of stock as payment for his efforts. The better Fitch got to know Voigt, the more he respected him. "He is a man of superior mechanical abilities and very considerable natural philosophy. . . . I soon perceived his mechanical abilities to be much superior to mine." Voigt was born in Germany in 1738, making him about six years older than Fitch. As a young man, he worked at the Royal Mint of Saxony at Saxe Gotha. Historians believe he immigrated to Pennsylvania around 1760 and worked for David Rittenhouse as a clock and instrument maker. He helped Rittenhouse, who was also an astronomer, build at least one of his intricate orreries, mechanical models that show the relative positions and movements of bodies in the solar system.[20]

With Voigt's help, Fitch set out to build the first steam engine either man had ever seen. They gathered up their tools and began to work in a rented workshop in the Kensington area several blocks north of the city, near the shipyards on the Delaware River. In June, the two men tested a model steam engine that had a cylinder of one inch in diameter. This device proved to be too small, so they ordered the casting of a three-inch cylinder, which wasn't completed until August. This model had limited success. Although the workmanship wasn't fine enough for the engine to work efficiently, Fitch could see its potential.

Fitch certainly wasn't the first man to try to improve a Newcomen engine. More than twenty years earlier, in 1764, a young man named James Watt was working as an instrument maker at the University of Glasgow. One day he was asked to repair a Newcomen scale model used in the classrooms. He immediately noticed how inefficient it was. Watt observed that the cylinder would produce more power if it could remain hot rather than being alternately heated and cooled.

In 1765, after nearly a year of off-and-on work on the model, Watt had a brainstorm (oddly similar to Fitch's Sunday-walk inspiration twenty years later):

> It was in the Green of Glasgow. I had gone to take a walk on a fine Sabbath afternoon. I had entered the green by the gate at the foot of Charlotte Street—had passed the old washing house. I was thinking upon the engine . . . and had gone as far as the Herd's House, when the idea came into my mind that as steam was an elastic body it would rush into a vacuum, and if a communication was made between the cylinder and an exhausted vessel, it would rush into it and might be there condensed without cooling the cylinder. . . . I had not walked further than the golf-house when the whole thing was arranged in my mind.[21]

Watt went back to his workshop and added a separate cylinder that would remain cool to condense the steam. Now the main cylinder stayed hot, increasing the engine's power. He patented the idea in 1769 but didn't produce his first working engine until 1776 or 1777, partly because he was busy surveying and managing the construction of canals in Scotland during that period and had little time to work on his inventions.[22] His next improvement, not patented until 1782, was the double-acting cylinder.

One of Fitch's first drawings for a steam engine, made while he worked alone in Warminster in August 1785, reveals that he somehow had hit on— whether independently or not—Watt's first improvement. This drawing shows something he called a "cistern" that was in fact a separate condenser. By August 1786, Fitch had incorporated Watt's second improvement, the double-acting cylinder, either independently or, more likely, as a result of his conversations with John Hall or Benjamin Franklin, who knew Matthew Boulton and had lived in Europe during the years of Watt's successes.[23] Fitch noted that their experiments with the three-inch cylinder engine "fully convinced us that a steam engine might be worked both ways as well as one," by which he meant that steam would be supplied to both ends of the cylinder— first one side, then the other, driving the piston back and forth, and doubling its power.

Building a steam engine was not the only challenge Fitch and Voigt faced. They also had to come up with a suitable design for a boat and the best method for propelling it. Fitch, in his earliest notes, proposed using paddlewheels that were twelve feet in diameter. He had once seen a boat in Boston that used paddlewheels turned by a horse walking on a circular treadmill. But almost as soon as he started thinking about them, Fitch had concerns about their efficiency; he was undoubtedly influenced by Franklin's criticism of them. He soon abandoned the idea. Like Franklin, he was concerned that a paddle blade pushed the boat forward in one position only: when it moved backward through the water. At all other positions, the paddle dragged through the water, slowing the boat and wasting valuable power.

Fitch also abandoned his next idea, shown in his 1785 model boat, an endless chain of paddles mounted on one side, probably for the same reason. Fitch thought about hinging the paddles on some sort of tilting mechanism so that they could move freely with the water while submerged. Had he done this, he would have had immediate success, for that was how paddlewheels eventually became reasonably efficient propulsion devices.

Fitch had read in Franklin's "Maritime Observations," the paper he had gone to so much trouble to obtain, that jet propulsion, in which air or water is forced through a tube in the bottom of a boat and so pushes the boat forward, would work better than paddles. This was Bernoulli's idea fifty years earlier. Franklin suggested that the pump might be powered with a fire engine, by which he meant a steam engine.

Fitch, like Rumsey, was eager to win Franklin's approval. "When I came to town I came with a design of making my boat with trunks to force water out abaft," Fitch wrote. But Voigt talked him out of the idea, sensing instinctively and correctly that Bernoulli's jet propulsion scheme was even more wasteful of energy than paddlewheels. Instead, the two men devised a "screw of paddles." This was a rather complex design, of a type also proposed by Bernoulli, featuring two helical propellers. Coincidentally or not, Fitch's design strongly resembled the screw propeller John Stevens would use with limited success in one of his early steamboats nearly two decades later.[24]

After weeks of work, one late summer morning Fitch and Voigt mounted their propeller device onto a small skiff and tested it manually, using a hand crank. Pushing off from the dock on the Delaware, the two men spent several hours trying to get the twin propellers to move the boat. By the end of the day, exhausted and dejected, they drifted back to shore. "[Voigt's] spirits seemed to be much depressed and as soon as we landed, he stole off from me and left me alone to take care of our machinery and stand the scoffs and sneers of those who waited our arrival. It is true I felt truly distressed but to remedy that I made myself pretty free that evening with West India produce [rum] and of course got some sleep."

The next day, Fitch fretted that Voigt was about to abandon the project. He knew that without his friend's mechanical skills he would be lost. He spent the entire day trying to come up with an honorable way to tell his backers that he was quitting and that they wouldn't get their money back because it had been spent. That night he was unable to fall sleep. Sometime after midnight, "the idea struck me of cranks and paddles for rowing of a boat. And after considering it some time [I] was sure it would be the best way that a vessel could be propelled by oars."

Fitch recalled hearing the night watchman call out one o'clock. He got out of bed, lit a candle, and began to sketch out his idea. "And when I drew it I was still more delighted with it as it brightened my ideas of the scheme. And I was at that time equally elevated with the scheme as I was depressed before." Before sunrise, Fitch rushed to Voigt's house on North Second Street, rousing him awake with the details of his latest brainstorm. "This gave new spirit to the undertaking and a small crank to make an experiment on a skiff was immediately ordered." To their immense relief, it worked.

Fitch's crank and paddle design was a clever way to get around the problem of wasted energy in paddlewheels. Twelve paddles—oars, really—were mounted on a frame above the deck, attached to arms that were moved by a crank. The paddles dipped into the water, six at a time, almost vertically on both sides of the boat, in the manner of men rowing a canoe. Fitch went back to his company directors and received approval to build a full-size boat and engine.

The money ran out early in the construction process. Fitch, always averse to begging his shareholders for more funds, returned to the Pennsylvania Assembly and proposed that instead of giving him exclusive rights (which he was still waiting to hear about) that they lend him £150. After he repaid the loan, he would ask for rights.

"Permit me, gentlemen," Fitch argued, "to inform you of the prospects we have before us. Mr. Voigt and myself are sure that we can build an engine; nay, we're vain enough to believe we can make one as good as they can in Europe. . . . Could I by any means raise sufficient money, I would not ask it from the legislature; but there is such a strange infatuation in mankind, they would rather lay out their money in balloons [a reference to the first hot-air balloons in Europe earlier that decade] and fireworks, and be a pest to society, than to lay it out in something that would be of use to themselves and country." His request went to the same committee that was considering his petition for exclusive rights. Its members decided to send the request to the full assembly. By a vote of thirty-two to twenty-eight, the assembly refused to give Fitch the loan.

Fund-raising was by far the worst part of steamboat building for Fitch. "I had to invent a new scheme every week, if not every day, to raise money for the purpose. . . . Could money have been extracted from my limbs, amputations would have often taken place . . . rather than to make the demands which I have." In September he wrote a letter to Franklin offering to sell the model of his first engine to the American Philosophical Society, at cost or whatever sum its members felt it was worth.[25] Franklin never replied.

Although little progress was made that fall and winter, Fitch's spirits soared when word reached him that both Delaware and New York had granted him term monopolies for fourteen years. This good news revived the flagging interests of the shareholders. To allow himself more of a voice in decision-making, Fitch reorganized the company, giving each shareholder no more than three votes. He also submitted a drawing and a short article to a Philadelphia monthly publication that reported on the work of manufacturers and inventors, the *Columbian Magazine,* to serve as public notice of his invention and perhaps to stir up financial support.[26] This article may have prompted Noah Webster to visit Fitch at his workshop and examine the boat, an event Webster recalled in an article he wrote for another magazine many years later.[27]

By the spring of 1787, "[we] began our work with redoubled ardor." But problems with leaky valves and defective condensers continued to plague him. To save money, Fitch and Voigt tried making cylinder caps out of wood, which proved to be a disaster. The horizontal cylinder kept leaking, so they tore it out and replaced it with a vertical one. At one point, Voigt came up with an idea for a pipe condenser. It performed a little too well—"the engine worked so brisk we could not find steam to supply it." Now they had to build a new boiler to keep up with the new condenser. The costs kept mounting and the shareholders began complaining again. A few tried to come up with other uses for the engine, including one idea for a steam-powered gun.

In April, the shipbuilder delivered their boat, which measured forty-five feet long by twelve feet wide. As Fitch and Voigt built a three-and-a-half-ton brick furnace on the deck, the sneers and laughter from onlookers increased. In early May, they finished installing seven tons' worth of machinery, including a 500-gallon boiler. They named this vessel the *Perseverance.*

The steamy hot summer of 1787 in Philadelphia will forever be remembered in American history, but not for Fitch's steamboat. The Constitutional Convention convened in May with the goal of creating a document to replace the weak Articles of Confederation. As the delegates hammered out compromises in Independence Hall, Fitch and Voigt tested their odd-looking steamboat a few blocks away. One day in late August, Fitch invited the delegates to take a ride, and several of them accepted the invitation. Others watched from the shore.[28]

That summer, Fitch's nemesis at the American Philosophical Society, Francis Hopkinson, wrote a letter to Thomas Jefferson, then serving as minister to France, about the goings-on in town. "We abound with schemers and projectors. There is one Fitch who has been this twelve-month endeavoring to make a boat go forward with oars, worked by a steam engine. He has made several unsuccessful attempts, and spent much money in the project and has heated his imagination so as to be himself a steam engine. I have no doubt but that a boat may be urged forward by such means, but the enormous expense and complexity of the machine must prevent its coming into common use."[29]

By summer's end, Fitch knew he needed a bigger engine if his boat was to succeed commercially. The *Perseverance* moved at just four miles per hour with a full load upstream—not fast enough to beat the competition, the stagecoach. Fitch knew that any successful steamboat would have to make the trip from Philadelphia to Trenton in less than five hours, the time it took a stage to travel the route. Unfortunately for Fitch, the Delaware River was one of the worst possible waterways on which to start a steamboat service. It was deep enough and wide enough for sailing ships to move easily up and down, and its flat banks encouraged the building of relatively good roads. Stagecoach service between towns along the river in recent years had become frequent, profitable, and easy.

Determined to build a more powerful engine, Fitch tried hard to persuade the company to advance funds for an eighteen-inch cylinder. But about this time, all the shareholders but one "gave up and declined their subscriptions." At his wit's end, Fitch wrote his swan song, intended for publication in the Philadelphia newspapers. He began, "I acknowledge I was vain in undertaking a business which I knew nothing about, that has taken near a century to bring to perfection. I mean the steam engine, especially when it was to be applied to a different purpose from any heretofore use." He then went on to explain how his boat would work, "I beg leave to suggest the advantages it would be to the United States. I suppose a twelve-oared barge would make equal speed with a stage wagon. . . . Which would save a great expense of horseflesh and feed. . . . And where streams constantly tend one way, great advantages will accrue to inland navigation, and in particular to the Mississippi and Ohio Rivers, where the God of nature knew their banks could never be traversed with horses."

He went on to dismiss criticisms that such a boat would be too large and heavy and would need frequent repairs. He then asked his readers to envision ocean-crossing steamships: "But the grand and principal object must be on the Atlantic, which would soon overspread the wild forests of America with people, and make us the most opulent empire on earth. . . . Pardon me, generous public, for suggesting ideas that cannot be digested at this day, what opinion future ages will have of them time only will make manifest." Fitch continued in a dramatic fashion for several more paragraphs, then announced that he must quit his works, not for lack of confidence but for lack of money.

Before taking this masterpiece to the newspapers, he showed it to a few of his shareholders. It "had the desired effect" and he never had to publish it. Reminded of their crucial role in this monumental project, the members of the company pulled together. They agreed to pay for a new, larger cylinder and sent Fitch to Warwick Furnace, near his former home northwest of Philadelphia, to have it cast. On his trip back, Fitch ran into a man who had recently been in Virginia and who mentioned that someone named Rumsey was experimenting with a steamboat in Shepherdstown.

Fitch flew into action. He stopped briefly in Philadelphia before setting off for Richmond to apply for exclusive rights from Virginia. Those rights were important to both inventors because Virginia was the largest state, both in area and population, and because it had several navigable waterways. In addition, its western boundary was the Mississippi River, the grand prize for steamboat projectors. Fitch was able to present the assembly a solid case: His boat was entirely different from the boat Rumsey received Virginia rights for in 1784 in that his was powered by a steam engine; he could also claim that he had a working steamboat in Pennsylvania. Success came quickly. On November 7, 1787, Virginia awarded Fitch exclusive rights to operate steamboats in its waters for a period of fourteen years, with one catch: Within three years, Fitch had to be operating at least two boats of twenty tons' burden each on Virginia waters.

After this easy victory, Fitch stopped in Annapolis a few days later to petition Maryland for steamboat rights as well. That state had given Rumsey navigation rights three years earlier for a mechanical boat of general description. An assembly delegate, the former governor Thomas Johnson, who had earlier tried to help Fitch, was confused about the kind of boat that Rumsey had received rights for. Not wanting the state to award conflicting rights, he asked the committee deciding the matter to delay its decision until he could ask his friend Washington if the boat Rumsey had shown to him in Bath in 1784 (and for which he received Maryland navigation rights that year) was powered by steam, as Rumsey seemed to imply.[30] Fitch, anticipating a long delay, returned to Philadelphia.

By early 1788, Fitch was still desperate to find other sources of funding. At the end of his rope, he decided to go to New York and ask the Continental Congress (which was still meeting; the Constitution had not yet been ratified) to do for him what it did for Rumsey two years earlier: promise him a tract of western land, which presumably he would sell with a caveat to investors if and when he put a steamboat in operation.

While Fitch waited in New York for a quorum to form, a friend introduced him to J. Hector St. John de Crèvecoeur, then the French consul in New York and author of *Letters from an American Farmer,* a book about life in America that was popular in Europe. The Frenchman immediately became enamoured with the idea of steamboats and offered to help Fitch obtain support and rights in Europe. The two men began a warm friendship marked by frequent correspondence over the next several years.[31]

To reassure himself about the feasibility of steamboats, St. John de Crèvecoeur wrote a letter to Franklin seeking his opinion of Fitch's steamboat; he enclosed a statement from David Rittenhouse (who by then had succeeded

Franklin as president of the American Philosophical Society) praising Fitch's invention.[32] Franklin wrote back saying that his health prevented him from getting out to see the boat but that hearing that his esteemed colleague Rittenhouse spoke highly of it "gives me more favourable sentiments of it." Franklin added that he had two problems with steamboats in general: that steam was not a strong enough force to move boats upstream and that such complicated machinery would take up too much room and be too expensive to maintain, limiting profitability.[33] That same month, Franklin wrote to his friend Jean-Baptiste Le Roy at the Royal Society in Paris and told him about Fitch's steamboat experiments.[34]

Fitch remained in New York until March. By then, representatives from nine states had shown up in Congress, not enough by Fitch's count to assure him of success. With work to be done on the boat, he decided to leave and come back when more states were present. A few days after he returned to Philadelphia, Fitch received bad news. After several months of work, the Warwick Furnace notified him that it was having problems casting the new cylinder to be strong enough to hold the force of steam. Fitch and Voigt dropped everything. They figured that they could solve the problem by lining the cylinder with copper. They traveled to Warwick to pick up the cylinder, only to learn to their dismay that the workmen, upset with their defective product, had destroyed it a week earlier. With no time that season to start over on a new engine, Fitch's only chance was to try the old engine on a different boat.

Then, to make matters worse, "Rumsey arrived . . . with his wicked and invidious pamphlets."[35]

FOUR

THE WAR OF THE PAMPHLETS

*Machines may be made by which the largest ships, with only one man
steering them, will be moved faster than if they were filled with rowers;
wagons may be built which will move with incredible speed and without
the aid of beasts; flying machines can be constructed in which a man . . .
may beat the air with wings like a bird.*

—*Roger Bacon, c. 1260*

Throughout 1787, Rumsey continued to stew as he heard about
Fitch's successes in gaining steamboat monopolies in several states.
He must have also learned about Fitch's steamboat experiments on
the Delaware that summer, although he never mentioned it in writing. For
most of that year, Rumsey and Barnes toiled in their Shepherdstown, Virginia, workshop rebuilding their ice-damaged boat.[1] It seems unlikely that
he was still a partner in the Bath boardinghouse or continued to operate his
mills; a deposition made later that year by his brother-in-law Charles Morrow stated that Rumsey had "for several years steadily pursued his boat
scheme, . . . to the total neglect of every other kind of business."[2]

In September, ten months after their first trial, Rumsey and Barnes were ready to try another test run on the Potomac. To simulate cargo, they loaded the boat with two tons of rock, fired up the boiler, and held their breath as the boat slowly gathered up enough power to move upstream—which it did, at about two miles an hour. But within minutes steam began escaping from several joints in the pipe boiler. They had to paddle back to shore.

Rumsey was disappointed but not discouraged. For all its problems, the pipe boiler seemed amazingly efficient. Poor workmanship caused the leaks, and they could be fixed. The boiler's twisted design required only about 20 pints of water, which Rumsey believed produced more steam than 500 gallons in a pot boiler. The lighter weight of the machinery—about 700 or 800 pounds—and its small size also offered distinct advantages on any boat meant to take on passengers and cargo. Rumsey and Barnes removed the boiler and took it back to their shop.

Sometime in November 1787, the news reached Rumsey that Fitch had won steamboat rights in Virginia. He knew he could wait no longer. A few days later he announced that the first public trial of the steamboat would take place in Shepherdstown on December 3. He thought it would be risky—"We should not have come forth publicly until spring if it had not been for Mr. Fitch's stealing a march on me in Virginia," he later wrote to Washington.[3] Rumsey seemed to think he could fight Fitch's Virginia monopoly by arguing that the 1784 Virginia monopoly rights he received for his mechanical poleboat extended to any other kind of boat he might later build. Unveiling this new model to the public would surely convince the Virginia Assembly that steam was what he had had in mind all along.

The day of the trial dawned clear and relatively warm, and by late morning a large crowd of men, women, and children began to fill the tree-lined riverbanks below the town. Among the observers were local residents General Horatio Gates and General William Darke, both of Revolutionary War fame, and most of the town's business, political, and religious leaders.

At noon, Rumsey, his partner McMechen, and his brother-in-law Morrow boarded the forty-eight-foot-long boat and began stoking the fire that would heat the boiler. A short time later, the invited passengers—Mrs. Rumsey and several of the town's ladies—were escorted to benches in the front of the boat.

An undated drawing of the event shows one woman busy at her knitting, as if to prove she was completely unfazed about riding on a wooden boat that had a blazing furnace on board.

Rumsey pushed the boat away from the dock and continued to stoke the engine. When the boat reached the middle of the river, it slowly turned around and began to move upstream. A thin plume of smoke flowed out of the stack. The crowd began to cheer. To eighteenth-century eyes, the vessel must have seemed to move by magic. No oars, no sails, no setting poles propelled it; only a gurgle of large bubbles at the stern hinted at the jet of water that the engine was pumping through the tube in the boat's hull.

The sounds coming from this early steamboat were oddly quiet for such a primitive machine. It did not make the loud, chugging, splashing sounds of the later Mississippi steamboats. In fact, the only noise the observers could hear—other than the shouts of joy from its passengers and crew—was a sound like the beating of a kettle drum: a resonant, rhythmic thump-thump, thump-thump, loud enough to be heard several blocks away.[4] "I was standing next to General Gates," wrote townsman Henry Bedinger years later. "When she moved out and he saw her going off up the river against the current by the force of steam alone, he took off his hat and exclaimed, 'My God, she moves!'" And indeed she moved, for almost two hours up and down the Potomac that sunny late fall afternoon. By the time Rumsey guided the boat back to the ferry landing, he must have been beside himself. One eyewitness wrote, "His own exultations left no doubt upon the minds of all that he had accomplished all his expectations."

Rumsey was ecstatic with this first success, even though he estimated the boat went just three miles an hour. On December 11, he made a second run, and by all accounts it went even better than the first. A few of the boiler's pipes sprang leaks, probably because water had frozen in them a few nights before, but Rumsey tied rags around the joints and still claimed to make four miles an hour against the current. Satisfied, he pulled the boat and engine in for the winter.

Within days of that last trial, Rumsey heard that Fitch had asked for exclusive steam navigation rights from the state of Maryland. Rumsey rushed to Annapolis to try to convince the Maryland delegates that steam was in his plans back when he applied for rights in 1784, long before Fitch thought of building steamboats. By then Washington had answered Thomas Johnson's letter to him earlier that month: "Mr. Rumsey has given you an uncandid account of his explanation to me, of the principle on which his boat was to be propelled against [the] stream. At the time he exhibited his model, and obtained [my] certificate I had no reason to believe that the use of steam was contemplated by him, sure I am it was not mentioned. . . . He then spoke of the effect of steam and the conviction he was under of the usefulness of its application for the purpose of inland navigation; but I did not conceive, nor have I done so at any moment since, that it was suggested as part of his original plan, but rather as the ebullition of his genius."[5] But in the end, the Maryland delegates denied Fitch's request. Rumsey apparently convinced them that being the first to *think* about steam gave him legal rights, even though his applications for monopolies to the various states never mentioned steam. It also probably didn't hurt that Rumsey was originally from Maryland.

Despite his recent successes, Rumsey was not sure how he could continue with his experiments. His frustration was evident in a December 17 letter he wrote to Washington from Annapolis. He told the general about his two successful runs on the Potomac, even enclosing copies of notarized certificates from several witnesses, but mostly he described the hardships he had endured and the threat that Fitch posed to his hard work. "[With] a large family to support, no business going on, indebted, and what little money I could rake together expended, a gentleman has since assisted me to whom I have mortgaged a few family Negroes which must soon go if I do not raise the money for him before long."[6]

As much as Rumsey dreaded facing Fitch head-on, he knew he had to leave the Virginia backcountry and continue his work in Philadelphia, where financial backers and skilled mechanics would be easier to find. He decided to pave the way by writing a treatise that would explain his boat and engine and at the same time launch a preemptive strike against Fitch. In January, he published a pamphlet in Shepherdstown titled "A Plan Wherein the Power of Steam is Fully Shewn by a New Constructed Machine for Propelling Boats or Vessels of Any Burthen Against the Most Rapid Streams or Rivers with Great Velocity."[7]

Rumsey began with a tribute to inventors and the burdens they must bear. "Those who have had the good fortune to discover a new machine . . . must lay their account to encounter innumerable difficulties; to correct a thousand imperfections (which the trying hand of experience can alone point out) to endure the smarting shafts of wit, and, what is perhaps most intolerable . . . to bear up against the heavy abuse and bitter scoffs of ill-natured ignorance. . . . Happy for him if he escapes with so gentle an appellative as that of a madman. . . . This is the fate of the unlucky projector, even in the cities of Europe."

Rumsey went on to claim that Fitch must have known about his plans for a steamboat as early as 1784 from a talkative Shepherdstown friend, Henry Bedinger, who traveled through Kentucky that year. Rumsey evidently knew about Fitch's surveying trips there, but what he didn't know was that Fitch's last trip to Kentucky was in 1781. In fact, it appears that Fitch did not know Rumsey was working on steam engines until the fall of 1786, well after he had approached several state legislatures seeking steamboat rights.

Rumsey then accused Fitch of twisting the truth, arguing that Fitch purposely misled the state assemblies that had previously granted Rumsey rights by "uncandidly" stating that his boat was completely different. Rumsey strongly believed that the rights granted to his mechanical boat also included rights to further improvements to it, including the use of steam as the source of power.

Rumsey justified his secretive approach by claiming that if he went public with his ideas, Fitch or some other unethical character would have stolen them. In the same paragraph, he states that his experiments revealed that a "boat might be constructed, as to be propelled, through the water at the rate of ten miles per hour, by the force of steam." He then says, as if to imply his boat had already achieved this speed, "Such a machine I promised to prepare, and such a boat to exhibit; this I have so far performed, in the presence of many witnesses." But the fastest speed Rumsey attained on either of his Potomac trials was barely four miles an hour, a fact he states later in the pamphlet.

Rumsey then twice accuses Fitch of stating that Rumsey stole his ideas of steam from him. He calls him a "dark assassin." He cites the technical superiority of his design, stressing repeatedly that his engine was simpler, lighter, and cheaper than Fitch's and that Fitch's engine was so heavy and poorly de-signed that it would never be able to move a boat against the current at more than three miles an hour. In all, the pamphlet contained four pages of argument and technical description followed by ten pages of sworn testimony from friends and neighbors.

Before leaving Shepherdstown for Philadelphia in late March, Rumsey wrote a letter to Washington explaining his plans for the future and enclosing five copies of his pamphlet. He did not mention that he had to sell the boat, minus the engine, to raise money for the trip.

> Tomorrow morning I shall throw myself upon the wide world in pursuit of my plans, being no longer able to proceed upon my own foundations, I shall bend my course for Philadelphia where I hope to have it in my power to convince a Franklin and a Rittenhouse of their utility by actual experiment, as Mr. Barnes is to set out in about ten days after me with all the machinery in a wagon and halt at Baltimore till I write him from Philadelphia what encouragement we may expect there, if none we will push immediately for South Carolina.
>
> There is no period in my life that could give me more satisfaction than to have it in my power to stop the mouths of the envious few (I might add ignorant) that has taken the liberty to cast reflections on the gentlemen that was so kind enough to give me certificates; one of this description would have got roughly handled by the gentlemen of this place if he had not made a very timely escape.[8]

The last few lines refer not to Fitch but to another victim of Rumsey's pamphlet, a Baltimore man named Englehart Cruse. As Cruse tells the story, a year earlier, in the spring of 1787, he was tending to some legal matters in Shepherdstown when Rumsey met him and invited him to come to his house; he wanted to show Cruse a picture of a steam engine in his copy of Desaguliers's book. Rumsey then invited Cruse to his workshop, where he showed him the cylinder he was constructing. It is unclear from Cruse's writing whether he had ever entertained the idea of building a steam engine prior to meeting Rumsey, but on his return to Baltimore, Cruse designed a stationary steam engine for use in raising water in a mill. He later asked the Maryland Assembly to grant him exclusive rights for the device. The delegates, who by then must have known about Rumsey's steam experiments, refused, probably afraid of treading on the 1784 rights they had awarded him for his broadly stated boat ideas.

Word of Cruse's attempt to gain rights soon reached Rumsey, who was sure that Cruse had "surreptitiously" stolen his ideas. In a second printing of his pamphlet in Philadelphia in the spring of 1788, Rumsey lashed out at Cruse and his "audacity." Cruse responded immediately with a fourteen-page pamphlet of his own, titled "The Projector Detected," in which he sarcastically tore into Rumsey's accusations and sharply criticized his steamboat design, claiming it would never work against a strong current. He also accused Rumsey of character assassination: "Mr. Rumsey!—You have given yourself the liberty to accuse me of such atrocious crimes, and to assert such a number of falsehoods, that I am really astonished, (as you say you are) to what latitude of ignorance, and to what longitude the current of self-conceit, has driven you from the point of truth and veracity."[9]

On his arrival in Philadelphia that early spring of 1788, Rumsey launched a whirlwind campaign to raise money. Dressed for success, Rumsey exuded charm and good manners. To most who met him, this backwoods inventor was a pleasant surprise, not at all like that madman Fitch, the laughingstock of the waterfront. Rumsey surely used his certificate from Washington to great effect when speaking to potential investors, even though the general's warm words were for a completely different kind of boat.

Rumsey's first and most important visit was to Benjamin Franklin, who must have been thrilled to learn that Rumsey had put his theory of jet propulsion into practice. He quickly bought one share in Rumsey's company. After that endorsement, it wasn't hard to sign up eighteen more influential Philadelphians. One man, Benjamin Wynkoop, was so enamored with Rumsey's plan he bought twenty shares. On May 9, the group of investors known as the Rumseian Society was officially formed. Franklin's name topped the list of shareholders. Around that time, Rumsey was admitted as a member of the American Philosophical Society, an honor Fitch would never receive.[10]

While Rumsey was busy forming his company that spring, another storm developed. The minutes of the April 18 meeting of the APS noted that the society had received papers announcing similar inventions from two men: James Rumsey and Henry Voigt. In addition to plans for improvements to mills and the use of a steam engine to raise water, Rumsey's paper described his pipe boiler for a steam engine. Voigt's drawings and descriptions were for various improvements to boilers for steam engines, including a pipe boiler.[11] The two boilers, except for the diameter of their pipes, looked virtually the same.

Rumsey immediately accused Fitch and Voigt of stealing his invention. He claimed that a Shepherdstown friend, a man named Askew, had talked with Voigt in Philadelphia several months earlier and extolled the benefits of Rumsey's innovative pipe boiler.[12] It wasn't long before a rumor began flying about town that Fitch had sent Voigt's pipe boiler plans to Matthew Boulton, James Watt's business partner in England, in hopes of getting his help in obtaining an English patent for the design.[13]

There is no evidence that Fitch ever did this, although around that time he did accept St. John de Crèvecoeur's offer to help him apply for French patents. Fitch also wrote to Jefferson asking for his help in securing French patents, but a reply from Jefferson has never been found.[14] This latest news of supposed Fitch treachery lit a fire under Rumsey's supporters. The Rumseian Society voted to send Rumsey to England immediately in hopes of beating Fitch to the patent office. While there, he could buy a steam engine from Boulton & Watt and build a steamboat in London. It all seemed so easy and made so much sense. English mechanics, the investors were certain, possessed skills and tools far more advanced than any in America. In addition, obtaining foreign patents would assure them of even greater profits in the future. Various shareholders penned letters of introduction for Rumsey to take to influential friends and colleagues living in London.

Fitch, in the meantime, spent much of April preparing a response to Rumsey's pamphlet. Fitch became aware of it about a week after returning from New York in late March. He had dropped by the house of one of his staunchest Steamboat Company shareholders, a prominent Philadelphian named Richard Wells, who was a former APS officer and an inventor in his own right.[15] Wells met Fitch at the door with a steely stare. As Fitch recalled later, "I was received as if he suspected me of fraud." Wells handed him a copy of Rumsey's pamphlet, which he said had been delivered by Rumsey himself.

Caught off guard, Fitch found himself tongue-tied ("I am so unfortunate that I am not able to communicate my ideas by word of mouth") and asked that Wells give him a few days "to set our matters right to his satisfaction." Fitch had great respect for Wells—"perhaps not a thousand on Earth to be found that can equal him in delicate sensations of honour"—and was determined to regain his trust.

He spent the next several days at his desk in his boardinghouse room writing a vigorous defense, attacking Rumsey's pamphlet paragraph by paragraph. He gave his draft to Wells, who read it and became convinced of Fitch's innocence. Feeling bad that he had mistrusted Fitch, Wells offered to edit the inventor's unrestrained ramblings into pamphlet form, a favor Fitch gratefully accepted.[16]

While Wells smoothed out the text, Fitch traveled—evidently on foot—to Frederick and Baltimore to get affidavits from workmen and others who had supplied Rumsey with parts for his steam engine and boiler. He hoped to counter Rumsey's claim that he had a working steamboat in late 1785. Fitch combed both places with the zeal and intensity of a modern-day investigator, searching out anyone who had ever cast a pipe or valve for a Rumsey machine. He gathered signed statements noting dates and circumstances from men named Tombough, Baltzel, Morris, Peters (all signed on April 17 or 18), Moale, Minshall, Raborg, Causten, Weir, and Zimmers (all signed on April 26 or 29).[17]

For example, the coppersmith who made some of Rumsey's parts said that it was March 1786 that Rumsey came and not before. A workman certified that he turned a piece of wood for Rumsey's boat in March 1786. A tinsmith for Rumsey's boat wrote that he began working on the parts in March 1786. An apprentice to the man (since deceased) who made copper parts said the work was begun in the spring of 1786. At first there was a problem in determining when Barnes ordered four brass cocks made in Baltimore because the foundry's account books had been destroyed, but Fitch kept at it until he found a worker who had some loose papers showing that the cocks were made on March 29, 1786. (Rumsey would later dispute these dates.)

Fitch ran into Barnes while he was in Baltimore and reported that Barnes had taunted him with the fact that Rumsey "exulted in the friendship of Washington, Johnson, and others, and went so far as to say that the one who could make the most friends would fare the best."[18]

"The foregoing testimonies," Fitch would write in his pamphlet, "I presume will carry full conviction that Mr. Rumsey has shifted his dates and has got two of his workmen to swear to it for Messrs. Barnes and Morrow, if they had consulted their accounts must have found that they had made a lapse of a whole year at least, and that the December 1785 they speak of must have been December 1786."[19]

By early May, Fitch's pamphlet was completed. But before he took it to a printer, Fitch did a bold and amazing thing. He paid an unannounced visit to Rumsey's Philadelphia boardinghouse room. Answering the door, Rumsey must have stepped back in shock at the sight of the rough-looking Fitch, eyes afire, papers clutched tightly in one hand. Introductions must have been unnecessary; Rumsey would have surely known who this agitated stranger was. It was the first and only time the two men met.

I have come to answer your charges against me, Fitch probably announced. *And, as a courtesy, to let you hear, before it is printed, my response.* Fitch then began to read aloud to Rumsey, word for word, the entire text of his pamphlet. Neither man ever recorded what happened that day, but since it appears no sheriff or doctor had to be called, the confrontation was probably tense but civil. Most likely, the well-mannered Rumsey figuratively sat on his hands and bit his tongue. Fitch read for half an hour, then abruptly left, no doubt slamming the door behind him.[20]

Fitch's pamphlet was printed in Philadelphia on May 10, the day after the Rumseian Society was officially formed. He titled it "The Original Steamboat Supported; or, a Reply to Mr. James Rumsey's Pamphlet Shewing the True Priority of John Fitch and the False Datings, Etc. of James Rumsey." Fitch wrote in the introduction that Rumsey's pamphlet charged him as "the perpetrator of crimes atrocious in their nature, but of which my conscience fully acquits me." He acknowledged that Rumsey might be a better engineer, but then followed that statement with a sarcastic blast: " . . . in the article of condensation I freely acknowledge he is my superior, having acquired the art of condensing (with the dash of his pen) one whole year into the compass of six days."[21]

Fitch admits the idea of steam power was not original with him. He mentions that William Henry of Lancaster told him in the summer of 1785 that he had thought about steam as a means to propel boats at least ten years earlier, and in fact had discussed the possibility at that time with Andrew Ellicott. He reports that Henry also told him that Thomas Paine had suggested steam-powered vessels in 1778. Citing Henry and Paine as examples of men with prior steamboat ideas, Fitch declares, "If bare projection was sufficient to build a claim on, I have no doubt but there are people now in the graves whose heirs might set up more early claims than either of us." The pamphlet continues with a chronological account of his progress over the next three years.

Fitch then tore into Rumsey's claims. Referring to Rumsey's 1784 demonstration of the poleboat in Bath to General Washington, Fitch offers proof that Rumsey never mentioned steam, stating that he merely alluded to it in a 1785 letter to Washington. Fitch even discarded Rumsey's claim to originality regarding the mechanized poleboat, saying that a similar boat "was many years ago tried on the river Schuylkill by a farmer near Reading, but without success." He noted that none of Rumsey's petitions to various state assemblies ever mentioned steam or even suggested the idea.

Fitch then asks why, if his ideas for a steamboat were as advanced as he claimed, Rumsey never objected when Virginia, Maryland, and Pennsylvania publicly awarded Fitch rights for his steamboat. Why, Fitch asks, would Rumsey not apply for rights for the use of steam on boats? "The reason is plain—General Washington gives it for him; it was an *immatured idea* and on which he thought he did not rely," quoting from the letter Washington wrote to Thomas Johnson.

He then wonders how Rumsey could say he "made considerable progress" in steam engines in the fall or winter of 1784 and then admit that he didn't begin to have parts made for one until October or November of 1785. Fitch argued that Rumsey was lying when he said he had an engine ready to go in the fall of 1785 when in fact the year was 1786. Rumsey said in his pamphlet that he was rushed into action in 1785 by a letter to his partner, McMechen, from a man named Buckley "living near Philadelphia" who told them about Fitch's experiments in that city. Fitch countered with the fact that he didn't move to Philadelphia until April 1786. To further prove his point, Fitch went to Lancaster and got a statement from Buckley stating that he wrote the letter to McMechen at the time when a Samuel Briggs was making patterns for Fitch's castings. Fitch got a statement from Briggs saying that he made patterns for Fitch in the summer of 1786.

Fitch defends his claim to the pipe boiler, saying that anyone experimenting with steam on a boat soon realizes that the size and weight of the furnace immediately sets the mind to figuring out a more efficient way to produce steam. He notes that although Voigt and Rumsey submitted their inventions to the APS on the same day, Voigt had "made a prior entry of his plans in the Prothonotorary's office in this city." To be fair, though, Voigt did this only when he heard that Rumsey was coming to town with his version. Fitch claimed that his earliest drawings contained a pipe boiler and that he had shown these drawings to Thomas Johnson in the summer of 1785 and submitted them to Congress that August. Fitch assumed that Johnson told Rumsey about the idea when he came to Johnson's foundry to have parts made that fall. Fitch explained that although the idea was original to him, he never put a pipe boiler to use because he feared that the twisted tubes would leak. (That spring, though, Voigt insisted on building a pipe boiler immediately, arguing that if he didn't, Rumsey would have a better chance of gaining rights to the device.)

Fitch notes in his pamphlet how silly Rumsey was to keep everything so secret and wonders what might have happened if Rumsey had publicly announced his ideas when he said he had them. Fitch reminds his readers that he had presented his plans to the Continental Congress well before the date that Rumsey ordered his first cylinder.

He tosses out Rumsey's argument that Rumsey's steamboat was technically better than his: "However faulty my works might be and however perfect his own, it would have no force in the determination of our title to the invention." Fitch closes by arguing for laws to protect inventors and give them the full benefits of their inventions:

The inventor can claim no benefit from his thoughts or inventions, before he makes a public declaration of such invention in some place of record established for such purposes; that is, he who invented and published a steam-engine will have an exclusive right for a certain number of years for all steam engines; at the expiration of which each improver has an undoubted right to the benefit of any improvement. On these principles, he who first invented and published the idea of a steam-boat, invests himself with a fair and just title to all steam-boats for a certain time, which in justice and policy government is bound to support.

By the time Fitch's pamphlet hit the streets, Rumsey was packing up to leave for London. He turned over the task of preparing a response to Fitch's pamphlet to his brother-in-law Joseph Barnes, instructing him to have it published as soon as possible. In the meantime, Rumsey made a few minor corrections to his original pamphlet, retitled it "A Short Treatise on the Application of Steam," and had more copies printed.

Rumsey had just a few days to wrap up his affairs before his ship left Philadelphia. On May 14, he wrote to his brother-in-law Charles Morrow in Shepherdstown and described his excitement, and his concern, over all that had happened in recent weeks. "This, Charles, is my meridian; if I do not do something now I am done." He continues, "I beg sir that you will leave no stone unturned to detect Fitch in his villainy, you shall have one of his pamphlets sent you as soon as they come out you can then judge what sort of proofs is wanted and forward them to our secretary [at the Rumseian Society]."

Rumsey's letter to Morrow closed with instructions of a domestic nature. "Charles, take care of my child and all the little business I left with you. . . . Remember me to Polly [his niece, Joseph Barnes's daughter by Rumsey's sister], tell her I sincerely wish her all happiness and tell my child [a reference to Susanna, his daughter from his first marriage] to be a good girl and remind her that it is in part for her I toil. Keep Neddy with the doctor if possible or some other school. I shall endeavor to have him some clothing got against winter and if nothing else can be done send him here to Joseph Barnes. I have laid a train for him to finish his studying but it will be expensive and therefore must be the last shift." (Neddy was Rumsey's son,

James, Jr., who lost his hearing after a bout with scarlet fever.)[22] The next day Rumsey was still at his desk. He wrote his will, leaving one-third of his estate to "my affectionate wife," and the remainder to his three children and a nephew.[23]

That same day he wrote to Washington, bringing him up to date on his steamboat progress and telling him about his move to London. He mentioned that friends had advised him to try to form a partnership with Fitch. Several meetings evidently took place between Rumsey and representatives from Fitch's company, but they rejected Rumsey's offer of an equal share partnership. The best the Fitch people could come up with, Rumsey reported, was to give him one-eighth of the company for his partnership, an offer he indignantly turned down, at which point "all negotiation ceased." He closed the letter with a request: "If you think, sir, that you could with propriety mention me in a line the first opportunity to the Marquis La Fayette, Mr. Jefferson, or any other gentlemen that you may think proper the favor shall always be gratefully remembered."[24]

The progress of steamboats certainly would have moved forward if these two bitter rivals had been able to pool their talents and financial support. But human nature being what it is, the steamboat wars chugged along. A few months after Rumsey's departure for Europe, Barnes and the Rumseian Society began appealing to state legislatures in Pennsylvania, Virginia, New York, and New Jersey to get Fitch's steamboat rights revoked. They had little success; the legislatures were not easily swayed to revoke rights they had already granted unless the reasons were compelling. Nevertheless, defending these claims drew Fitch away from steamboat work.

In New York, a third party entered into the fray for navigation rights: John Stevens, the wealthy businessman who had reneged on his promise to help Fitch in New Jersey two years earlier. Stevens, his ideas only on paper, proposed a steamboat that used jet propulsion, the power for which would be created by a pump operated, rather amazingly, by an inefficient and bulky Savery engine. Stevens argued that his plan was different from the plans of both Fitch and Rumsey, but the New York Assembly denied his request.[25]

In July, Barnes published Rumsey's second pamphlet, "Remarks on Mr. John Fitch's Reply to Mr. James Rumsey's Pamphlet," in which he took issue with Fitch's accusations of date switching. He explained that the confusion arose from the fact that Rumsey had parts made for steam engines at different times—both in the fall of 1785 and during 1786—and that Fitch chose to provide depositions for the 1785 parts only. Barnes then accused Fitch of bribing workmen to obtain false affidavits. He printed the statement of one such man, Charles Weir, who said he had told Fitch that he wasn't sure exactly when he had made brass cocks for Rumsey, and that Fitch in reply "persisted in soliciting me, and said he would give me anything I would ask, only to prove it, and I should be handsomely rewarded; but I positively refused." He added that he saw Fitch a month or two later and told him he had found the receipt, but that Fitch's only concern then was that Rumsey's people would probably make an issue of it. Barnes was also able to get an affidavit from another worker who stated that Fitch had bribed one of the apprentices in their Baltimore shop to make a favorable deposition. In all, Barnes offered nineteen affidavits to contradict Fitch's claims.[26]

These charges further damaged Fitch's reputation. Rumsey's first pamphlet had already doomed his second attempt to get support, in the form of a land grant, from the Continental Congress.[27] Fitch had left his petition with friends in New York, which they were to submit when conditions appeared favorable; it had to be withdrawn because of the bad publicity resulting from Rumsey's pamphlet. "When I received information of that, and reflecting how I had ruined myself to serve my country . . . it affected me beyond measure," Fitch recounted.[28]

While Fitch spent much of the spring and summer of 1788 battling Rumsey, Voigt continued to work on their steamboat. The biggest challenge was to develop a leak-proof pipe boiler and fit it in their new, longer boat. Fitch's plan to install an eighteen-inch cylinder had been dashed when the foundry destroyed it out of frustration earlier that year. With no time left to cast a new cylinder before winter, Fitch and Voigt decided to build a new boat, longer and narrower than their first boat, thinking that "a 12-inch cylinder might move that with the same velocity as an 18-inch cylinder would the other."[29] With its high length-to-beam ratio of sixty feet long and eight feet wide, the new boat would slip more easily through the water than the forty-five-by-twelve-foot model.

They also changed the way in which the oars were mounted. Instead of hanging them off the sides of the boat, they moved them to the stern. Rather than twelve oars, they used three or four large paddles shaped "like snow shovels." The weight of the boat was lightened considerably when they installed the new pipe boiler, since they could then remove three and a half tons of furnace brickwork.[30]

"After experiencing innumerable difficulties more than any mortal man can conceive, except Mr. Voigt and myself, we got the boat to work tolerably well and set off on a journey to Burlington," about twenty miles upriver from Philadelphia. The voyage, made in July, was the longest nonstop trip made anywhere by a steamboat to that date. They took along as passengers two of their most enthusiastic shareholders, Richard Wells and the physician Benjamin Say. More than half a century later, an account of this first long voyage—and it can't be known whether the scenes described are accurate or imagined—appeared in the July 1855 edition of the *Trenton Gazette*:

> Crowds of persons assembled at all the prominent points along the river to see her pass, and waited for hours to witness what was then the greatest wonder of the day. At Point-No-Point, now Bridesburg, the whole population of Frankord and the upper end of Philadelphia County were assembled, and they saw the boat slowly steam by them on her upward progress. Great indeed was their enthusiasm, and long and loudly they did cheer the grotesque exhibition. Women waved their handkerchiefs in approbation, bateaux put off from shore and rowed alongside the steamer cheering the adventurous and now exulting Fitch. At Dunks Ferry a similar demonstration took place as the new boat steamed onward. A vast concourse of people had collected there from the interior of Bucks County to witness the passing of the new wonder.
>
> At length she approached her destination. So far everything had gone on to the satisfaction of Fitch, whose crudely constructed machinery had performed its office for several hours in succession without any faltering. The green bank at Burlington was thronged with ladies. . . . The town wharf was also densely thronged with people. As the boat came opposite the wharf she rounded to, and even while the cheering went up, both long and loud, she unexpectedly dropped anchor in the middle of the river. A batteau was sent off to learn the cause, when it was discovered she had burst her boiler![31]

Fitch's own account of the trip focused not on the cheers from along the shore but on the embarrassment near the end. "Our cussed pipe boiler got such a leak we could not work the engine any further." They dropped anchor, and Wells and Dr. Say yelled back to the boatsmen who were shouting taunts at them that they had dropped anchor on purpose. But Fitch couldn't abide such a lame excuse and shouted the truth. Unable to make a repair on the spot, Fitch let the boat float toward the Pennsylvania shore, where the passengers disembarked to find a quicker way back to Philadelphia. Fitch and Voigt drifted back with the tide during the night. They were not discouraged, though; "we knew that if we could once make our boiler tight" the boat would be a success.[32]

They removed the boiler and worked on it for several more weeks. On October 12 they set out for Burlington once more, making the journey in three hours and ten minutes. Fitch invited supporters and influential citizens to come along on that trip and several others that he made that month. Many of his passengers gave him certificates of success, including prominent men such as Andrew Ellicott, John Ewing, and David Rittenhouse. But Fitch remained disappointed; the boat never attained the speeds he had hoped for: " . . . for although we had made our boat go fast enough to answer a valuable purpose on the Ohio, it did not go fast enough to answer a valuable purpose for stages on the Delaware. For unless it can be carried to Trenton which is 38 miles, in less than five hours, it will hardly be an object worth attending to on the Delaware."[33]

Fitch knew that this news would cause his investors to back out. He could hardly blame them for being unwilling to continue funding a venture that was so unlikely to be profitable any time soon. But he was taken aback when Voigt abruptly quit that winter, claiming that working on the steamboat had caused his family life and watch-making business to suffer.

Nevertheless, Fitch continued to press on. In late 1788 he came up with the idea of forming a new company, issuing forty shares at $10 each, with the idea that when the new boat proved successful he would merge the two companies. To his surprise, a few of his original shareholders contributed to the new company. "These proposals, calculations and estimates soon set the scheme afloat again." Plans were made to try once more to cast a larger cylinder and put it on the boat the following spring.

One of Fitch's new supporters was an enthusiastic twenty-nine-year-old physician named William Thornton, who had purchased sixteen shares in the new company. Thornton, who became a U.S. citizen that year, was born in Tortola, in the British Virgin Islands. A man of many interests, he held a medical degree from the University of Aberdeen in Scotland. While working with Fitch he began his lifelong career as an amateur architect, winning a competition in 1789 to design a building for the Library Company of Philadelphia.[34]

Around that time, Fitch got permission from the shareholders to hire John Hall, the inventor whom Fitch had consulted two years earlier. Hall was living in Bordentown, New Jersey, at the time, working off and on for Thomas Paine on his iron bridge projects. Hall agreed to come to work for Fitch that winter.[35]

Although he still had high hopes, life wasn't easy for Fitch. Constables and workmen hunted him down to collect debts. Walking the streets of Philadelphia, when he dared, subjected him to taunts, even from former friends and supporters. He later wrote about this period, "But the former embarrassments, which I have mentioned, were but inconsiderable when compared with other matters considering the indignities offered me by my best friends and patrons, who in many instances treated me more like a slave than a freeman, whilst I was in the most excruciating tortures of devising plans of completing my undertaking which was far beyond my abilities."[36]

The state of his wardrobe, never good, was by then deplorable: "one pair of breeches with a break in the crotch, . . . one coat" of which "I cannot tell the exact age and constant wear of it, but it is broke in every part, especially the lining, elbows, about the wrists and under the arms. . . . Every decent man must and ought to despise me from my appearance." Except to eat and sleep, he spent little time at his boardinghouse, where "I was obliged to suffer just indignities from my landlord and be henpecked by the women. . . . Which always in the evening drove me to a tavern."[37]

FIVE

THE *Columbian Maid*

Which do we receive the greatest benefit from: our friends or enemies, as to useful lessons in life?

—*John Fitch*

The Sign of the Buck Tavern, on Second Street between Vine and Race, had been a pleasant retreat for Fitch and Voigt for the past two years. There they escaped the taunts and jeers of the waterfront and relaxed with the amiable owner, Michael Krafft, and his wife, Mary. When Michael died, Mary continued to run the tavern to support herself and her several children. Soon Voigt, Fitch, and Mrs. Krafft settled into a comfortable nightly arrangement. Fitch called the other two his only true friends in the world.

Fitch would leave the tavern each night around eight o'clock to retire to an early bed at his boardinghouse a few blocks away. After a year or so, Fitch noticed a change between Voigt and Mrs. Krafft, as he referred to her. "I began to suspect their friendship was carried to excess." His suspicions were confirmed when sometime in the spring of 1789, "the productions of love appeared," as Fitch obliquely referred to a visibly pregnant Mrs. Krafft.[1]

Fitch was appalled. He took aside the married Voigt, father of six or seven children, and lectured him on the destruction that was about to befall both good families. He had agonized over his friends' situation for a long time, he explained to Voigt, and had come up with a solution. "I never design or intend to marry," he told his friend, "therefore what I am about to propose can be no inconvenience to me. And to save your and her reputation, I will marry the woman and pledge my word of honor never to bed her." Voigt thought this was a fine idea, but when they took it to Mary she viewed it "with considerable resentment"—even though, as Fitch heartily insisted, that "there was nothing but the purest and undefiled friendship designed."[2]

Mary eventually had to close the tavern and shut herself inside; pregnant widows were not highly regarded in Philadelphia society. Fitch stayed close by—"I did not or could not forsake her"—for several weeks before the birth and a month after. Voigt stopped coming by. As a result, Fitch had to endure "the griefs and hysterics which it caused. And the sleepless nights which I experienced made my heart bleed for the woman's distress."

When the birth seemed imminent, Fitch hired a midwife who had assured him she would bring an assistant. But when the time came, she arrived alone and called on Fitch to help, to his considerable chagrin. "After which I was obliged to degrade the man and become a nurse." But he writes that even so, he was glad to help his friend.

The baby—a boy Mary named John Barney Voigt—arrived on a Sunday evening in August, in the middle of a violent storm. Fitch was thankful that the noise of the thunder drowned out the cries of the newborn infant and helped save Mary from revealing the shameful event to the neighbors. Later that evening Fitch bundled up the child and took him to a woman who had agreed to raise him, an act that made him quite uncomfortable—"This I look upon to be the greatest crime that I was ever guilty of."[3]

Not long after the baby's birth, Voigt returned to work with Fitch; surely the personal tensions between the two men had put stress on their steamboat partnership. Voigt also resumed his affair with Mary. Fitch did everything he could to cool their relationship. "By gentle lectures to each of them as often as opportunity served I religiously preached up the valuableness of their families. . . . They both frequently promised me that would comply with my advice."[4]

━■━■━■━■━■━■━■━

The Rumseian Society had failed in its first attempt, in September 1788, to have Fitch's exclusive rights for operating steamboats in Pennsylvania revoked. Their efforts to strip Fitch of rights in Virginia, New York, New Jersey, and Delaware also met with defeat that fall and winter. In March 1789, the society decided to try again in Pennsylvania, this time taking a different approach. Two Rumsey supporters, Philadelphia Federalist politicians Thomas Fitzsimmons and George Clymer, proposed that the state create a commission that would have the power to grant patents.

Fitch could see where this was leading and protested. The question of the legality of such a commission was presented to the Pennsylvania Supreme Court: "Can this House, consistent with the principles of law and justice and the Constitution of this State, enact a law upon the principles reported before this House, in the case contested between John Fitch and James Rumsey?"[5]

Lawyers for each side were hired—Miers Fisher for the Rumseian Society and Richard Wells for Fitch. Fitch and Wells presented a solid case, clearly developed and summed up in twenty-four well-argued points. Thompson Westcott, an early Fitch biographer, notes the judges' findings, announced in late March 1789:[6]

> . . . The opinion of Chief Justice McKean was evidently biased by prejudice. He argued that if Fitch's law was obtained by deception, it might be repealed. . . .
>
> Judge Bryan was opposed to disturbing the law. He referred to English laws granting monopolies, and showed that they had been sustained for the reason that, having been passed, it was better to abide by them than disturb the course of law, as it was a mischievous thing for government to have its faith suspected.
>
> A third judge gave no written opinion. The Committee, upon consultation, again reported in favor of Fitch, by declaring that the passing of a patent law by the State was inexpedient.[7]

A few days later, Fitch, even though he had won the case, was still bitter at the trouble it had caused him. He wrote an angry letter to the editor of the Philadelphia *Independent Gazetteer* accusing Fitzsimmons and Clymer of trying to take away his lawful rights and attempting to ruin him. "You are now going to Congress, and [I] wish to have it known to your fellow citizens, that I deem you my professed enemies on this subject, and that you will leave no stone unturned to hurt my interest with that honorable body."[8] This was not the end of the Rumseian efforts. Fitzsimmons submitted a bill later that spring to obtain rights for several of Rumsey's inventions, all of which the assembly approved except for boats propelled by steam, which of course did Rumsey no good in his fight with Fitch.

That May, with the Rumsey threat weighing on his mind, Fitch "decided to set off for Shepherdstown, where Rumsey did his mighty feats, in order to obtain the truth of his assertions." He had time on his hands because the Atsion Furnace in New Jersey, which had been hired to cast the new cylinder, had never taken on such a large job and was having problems; the delivery date had slipped from April to June.

Fitch arrived in Shepherdstown and took a room at Wynkoop's Tavern, using a false name. As he describes the visit in his autobiography, Fitch was forced to confide his true identity and business to the owner, Cornelius Wynkoop, in order to get information about Rumsey's boat. Fitch worried about that but hoped that Wynkoop would, in the interest of keeping a paying guest, keep his identity a secret. Fitch first asked Wynkoop how fast Rumsey's boat went. "As fast as I could walk," Wynkoop said. He quickly corrected himself to "Faster than I could walk" when he saw Fitch taking notes. Many other questions followed, and Wynkoop's answers confirmed Fitch's suspicions about the deficits of Rumsey's boat.[9]

Afterwards, Fitch hung around Wynkoop's bar for while, trying to interest his fellow drinkers in idle gossip about Rumsey. When he failed to arouse much comment, he left and headed for another tavern in town. There he picked up a few tidbits. One man told him that Rumsey's boat had never gone farther than four hundreds yards under steam. Another story involved a bet made between Rumsey's brother-in-law Barnes and a Captain Ross:

Barnes wagered that the boat would exceed three miles per hour. When it didn't, Ross collected on the bet.

Fitch returned to Wynkoop's Tavern to retire for the night. Wynkoop saw Fitch come in and pulled him aside, warning him in a low voice that he had heard that Charles Morrow, another Rumsey brother-in-law, had figured out who he was (no doubt Wynkoop had told him) and was planning to "abuse" Fitch if he could find him. In fact, Morrow was sitting in Wynkoop's bar at the time. After glaring icily at Fitch for a few minutes, Morrow walked over to the inventor and asked to speak with him in private. Fitch replied, "Mr. Morrow, I have no private discourse to hold with you," and moved farther into the crowded barroom. Morrow followed him in and began yelling curses and insults. Fitch took this abuse for a while, but when he noticed that the crowd was cheering for Morrow and laughing at him, he slipped out.

Fitch decided it was time to leave town. He crossed the Potomac and stopped for a while in the Maryland town of Sharpsburg. There he composed a note to Morrow proposing that they settle their differences in court, not in a barroom. Fitch continued crisscrossing the local back roads, gathering information on Rumsey from taverns and workshops in the area. He wasn't far from Fredericktown, where so much of Rumsey's parts making had taken place. Word traveled fast about Fitch's presence in the area, and the tall, gawky stranger was easily spotted. One evening, as he approached a tavern, he heard someone yell out a window, "Here he is!" A heavyset man hurried out to the tavern front door and, in the process of looking at Fitch, missed the porch step and fell flat on his face.

Fitch stepped over the man and walked into the bar. Seeing the angry stares of a roomful of drunken iron forge workers and woodchoppers, he decided to feign exhaustion. He found a couch and pretended to sleep—"the greatest ruffian could not attack a tired man lying on a bed," he reasoned. As insurance, before closing his eyes he ordered a round of whiskey for everyone. A few refused Fitch's offer, including the man who had fallen at the door. They ignored him for a while, until a supper of boiled potatoes and milk-bread was served. Fitch was hungry but refused the meal, afraid to sit up and face the crowd.

Out of a half-closed eye, Fitch watched the heavyset man—whose name was Crampton and who was deputy sheriff for Washington County, Maryland—walk over to the food table and pick out a large potato from a bowl. "I at once let in an opinion that it was intended for me, and turned myself . . . to watch him, and kept my eyes fixed on him. Some short time after, getting one of his companions between him and me, he threw the potato with all his seeming strength and hit me on my breast near the pit of my stomach, but the hurt was not great."[10]

Fitch rose from the couch and began berating Crampton for such a cowardly act. The drunken crowd was caught off guard by Fitch's outburst and grew quiet. Crampton and a few of his friends left, threatening to return the next morning. Taking no chances, Fitch checked out early and started the long walk back to Philadelphia. He stopped in Harpers Ferry on the way and was able to gather a few more affidavits about Rumsey's activities.

Back home, Fitch was encouraged by the progress John Hall had made in producing a new condenser. But he faced another setback in June. When the new cylinder was finally delivered, it was so crudely made that Fitch had to line it with copper sheeting. "We made engineers of common blacksmiths," Thornton lamented years later.[11] Forge workers were not skilled enough, nor did they have the proper tools to create parts with the high tolerances required by steam engines. It was the end of August before the machinery was ready to be put on the boat.

By then, Dr. Thornton had come up with a condenser of his own, made from a thin sheet of eight-pound copper. He insisted on trying it out, which meant the boat's workings had to be taken apart to remove Hall's condenser and install Thornton's. Fitch was beside himself with frustration. Not only did this create another delay, but he was sure that the new condenser would be too flimsy. *We'll try your design,* Fitch urged Thornton, *but let's use eighteen-pound copper, not eight.* But Thornton insisted, throwing the weight of his sixteen company shares behind him. Fitch gave in. Thornton's condenser was installed and tested. It immediately "crushed in like an eggshell." Only then did Thornton agree to try the design using heavier gauge copper.

While they waited for Thornton's new condenser to be made, Fitch reinstalled Hall's condenser and gave it a trial run. The engine worked well but moved the boat no faster than it had gone a year earlier. Whatever Hall knew about Watt steam engines, it wasn't enough to help Fitch.

Voigt became aware of Fitch's lack of progress that summer and returned in time to help install Thornton's stronger condenser—by Fitch's count, the seventh condenser to date. The result was the same as with Hall's condenser: The boat moved well but no faster than before, even though a much larger cylinder was now in place.

Fitch, frustrated and weary of having to beg his supporters for more money, gave up. He turned everything over to Voigt, who, refreshed by his several months' absence from the project, had many new ideas. The shareholders were reassured by Voigt's reappearance and told him to go ahead. His first idea was a new kind of pipe condenser, but when tested it also failed to increase the engine's power. He then came up with a way to force a jet of water into the condenser. To this end, he invented a "curious forcing" pump, but when it was tried, the result was "not a whit better."

By now winter was approaching and Fitch became involved once more. Where had they gone wrong? He went over every piece of machinery in his mind. He hit on the air pump and decided to enlarge it. That "brought the engine pretty nearly to perfection," he wrote. They gave it a second trial, when they noticed that the condensation was imperfect. Some alterations were made and a third trial was set.

Early in the morning on the appointed day, Fitch went down to the waterfront to get the boat ready. He started a fire under the boiler. Not long after, a sudden storm blew in. Strong winds convinced him to delay the trial. He put out the fire, but not completely. Late that night Fitch was roused out of bed by shouts from below his window that his boat was on fire. He frantically dressed and ran to the waterfront to see flames burning holes through the woodwork near the furnace. Somehow he was able to sink the boat to save it and his machinery from complete destruction.

In December, the boat was raised, patched, and tried again, to further disappointment. The engine worked well, but the speed remained the same. At no more than four or five miles an hour, it could never compete with stagecoaches along the Delaware. Fitch pulled in the steamboat for the winter. Dejected and broke, he walked home to Warminster to look for clock

repair work. When he returned ten days later, still unable to pay his landlord for room and board, Mary Krafft took him in. She assured him he
didn't have to pay her until he had the money.

■━■━■━■━■━■━■━■━■

Rumsey, in the meantime, had been living in England for eighteen months
by the winter of 1789–90. As soon as his ship docked in June 1788 he made
a beeline to Birmingham to visit Boulton & Watt. They must have been expecting him. One of the many letters written on Rumsey's behalf from
Philadelphia earlier that year was from the well-respected American physician Benjamin Rush to his English friend, the similarly respected London
physician John Coakley Lettsom:

> A certain Mr. Rumsey, of Virginia (strongly recommended by General Wash-
> ington) lately produced a plan of a machine in our city for improving the
> steam engine. . . . This plan, it is suspected, has been copied with a few tri-
> fling plagiarisms by a person in this city [meaning Fitch] (equally known for
> plagiarism in philosophy, and a licentious opposition to the proposed consti-
> tution of the United States) and transmitted to Mr. Boulton of London, with
> a view of obtaining a patent for it.
> The only design of this letter is to request you to suggest the above infor-
> mation to Mr. Boulton, and to assure him that the proper vouchers will be
> sent to him by Captain West or Captain Sutton in a few weeks, who will prove
> that the sole honour of the invention belongs to Mr. Rumsey; he alone is en-
> titled to it.[12]

Matthew Boulton and James Watt gave Rumsey a warm welcome. By
then the two men had been in partnership for twenty-three years, with Boulton providing the financial and management expertise that the inventor
Watt lacked. They were impressed with Rumsey and his list of distinguished
backers. After meeting off and on for two weeks, the three men came close
to forming a partnership to build steamboats.

"So far were we agreed by the twelfth [of July] that a petition was drawn
up for me by Mr. Watt, wherein I prayed his Majesty for a patent for propelling vessels of all kinds by steam, and a boiler on a new principle for
steam engines, expecting to gain both privileges with the expense of one

patent," Rumsey wrote William Bingham, who was president of the Rumseian Society back in Philadelphia.[13] He described how Boulton had notified his company attorney in London two days later to withdraw a caveat he and Watt had filed at the British patent office stating that they would oppose any application to patent steamboats. Rumsey waited around for several days while Boulton drew up a proposal to formalize their agreement.

When Rumsey read their offer, he was shocked at the proposed terms. As he told Bingham, " . . . they were so far from what I had expected, from the conversation that passed on the subject before, that it really astonished me." After briefly discussing his objections with them, Rumsey told Boulton that he would have to confer with his advisers in London: Benjamin Vaughan, an English businessman and old friend of Franklin's (they had worked together on the 1783 peace treaty negotiations in Paris), and Robert Barclay, a London banker who owned shares in the Rumseian Society.[14] Vaughan's brother, John, was a member of the Rumseian Society in Philadelphia; interestingly, their father, Samuel, had been an investor in Fitch's original Steamboat Company.[15]

Rumsey seemed to be most upset with the Boulton & Watt condition that required him to dissolve the Rumseian Society. Boulton, who believed that the American company would be unnecessary and a potential source of conflict, offered to provide Rumsey with the money to repay his investors. Rumsey, however, told Bingham that to do so would be "dishonorable, or incompatible with the engagements that I had entered into with gentlemen who had given friendly support to my enterprise."

But there were many other points of disagreement. When Rumsey got back to London, he showed the proposal to Barclay, who was "much displeased." Vaughan also convinced Rumsey that he could do better and encouraged him to negotiate on several points. In the meantime, Vaughan wrote to Jefferson, then U.S. minister to France, asking for advice in obtaining a French patent.[16] Jefferson replied later that month, telling Vaughan he had consulted with the French secretary of the academy of arts and sciences, who gave him instructions on how Rumsey should proceed.[17]

In early August, Rumsey wrote a long letter to Boulton and Watt citing his objections to their proposal.[18] First, he took issue with their seemingly magnanimous gesture to remove their patent claim to propelling boats by steam. Rumsey held that such an action would benefit him only in England. To their offer of providing steam engines to Rumsey "on the usual terms," Rumsey replied that this was no favor, especially in light of the next term, which stated that Rumsey must use only Boulton & Watt engines in any country in which he had a patent. For that concession, Boulton & Watt would pay half the cost of obtaining an English patent. Rumsey remarked, "This is a great return required for less than one hundred pounds purchase money."

Next, Boulton and Watt had asked that Rumsey take out a partnership patent in America for their engine and all its applications, for which they would pay two-thirds of the cost and receive two-thirds of the profit. Rumsey countered that although he had no problem if Boulton and Watt obtained an American patent for their double-acting engine, he would not let them take credit for and earn more than he would from his invention, the tubular boiler.

Finally, he objected to their requirement that he use only Boulton & Watt engines anywhere in the world. Such a term "deprives me of all means of prevailing myself of future improvements, even my own, except on your terms . . . and leaves me to be undermined by any rival." So as not to "tie my hands in other places," Rumsey countered by proposing that any agreement between them would apply to Great Britain only; that they would pay him one-half of the profit on steam engines used for navigation in England; that they would pay for half the cost of an English and Scottish patent and pay for his voyage back and forth to Europe; and that they would compensate him if they used his tubular boiler in any future engines.

Rumsey asked Vaughan to write a statement of his objections to the proposal, which Rumsey included with his reply. One of Vaughan's suggestions irked Boulton and Watt considerably. He told Rumsey that he didn't need Boulton and Watt and offered a way Rumsey could avoid a patent conflict with them: Rumsey could hire English mechanics to manufacture his steam engines in Ireland, a country in which Watt's patents did not apply. When Watt read this, he told Boulton that it "shows the principles of the man in so clear a point of view that I cannot express how much I detest him." Boulton called Vaughan "a dangerous man."[19]

Boulton sent a reply to Rumsey about a week later. "I have little hopes of our forming any connection, because you reason as if you expected shadows to counterbalance substance." Boulton admitted that selling engines to Rumsey at the regular price might not seem like a bargain, but he argued that their company deserved a fair profit since they were taking on the risks and expenses of manufacturing. He also explained that Rumsey was buying the many years of knowledge and experience that went into making the engine.[20]

As for Rumsey's boiler, he replied that when Rumsey "has tried as many boilers as we have—hundreds—he will have discovered some mistakes in it. There are many cases in which your boiler cannot be used and there are many in which it may and therefore we wish not to discourage you from taking out a patent for it and we'll render you any assistance in our power in the introduction of it." Boulton concluded, "Partnerships ought to be founded on equitable principles and like a pair of scales be balanced, with money, time, knowledge, abilities, or by possession of a market . . . now it appears to me that our sentiments are not in unison and that you have mistaken your road to the goal in view."[21]

But Rumsey wasn't ready to call it quits. He fired back a counteroffer on August 22. A week later, Boulton and Watt sent their final answer. "Though we are in general much averse to partnerships . . . yet from the very favourable opinion we formed of your abilities and character we were induced to make you offers which we shall probably not make to any other person in the like circumstances. . . . We do not think there is any prospect of any terms being proposed which could make a connection agreeable to both parties; it is therefore our opinion and desire that the negotiations should now terminate, having nothing further to propose on our grant, except our wish to remain on a friendly footing with you." They added that they would not interfere with Rumsey's attempts to patent his pipe boiler in England. But, they warned, they would hold on to their claim to the right to use steam engines in navigation, regardless of the method of propulsion, and they would fight any English patent application that he might file in that regard.[22]

Rumsey's overconfidence and sense of honor had done him in. He was so sure he was close to building a successful steam engine that he felt he was on equal footing with Watt, who had been working on steam engines for more than twenty-five years. Had Rumsey agreed to a partnership, he almost certainly would have had steamboats up and running within a year or two. Robert Fulton—whose first successful steamboat nearly two decades later relied on a Boulton & Watt engine—might not have made it into the history books.

Full of confidence, Rumsey pressed on. In November 1788 he obtained an English patent for his tubular boiler, steam engine, and steamboat. Evidently, Boulton and Watt did not make good on their threat to fight Rumsey on patent rights. Knowing that funds from the Rumseian Society would not be sufficient, Rumsey began to look for financial backers in London. He eventually came to terms with a man named Whiting, who offered to come up with the 600 guineas needed to build a boat.

In January 1789 he wrote to his brother-in-law Charles Morrow that things were going well. He reported that his patent applications were in place in Holland and France, and he had sufficient support to begin steamboat experiments that spring. "My intention is to get a vessel (if possible) large enough to go to France and Holland by steam alone."[23]

In March he wrote to his brother Edward that his boat was being built in Dover, and that it was "large enough to go to the East Indies," of one hundred tons burden, much larger than Fitch's boat. He noted that the engine was being built (he did not say by whom—presumably workmen under his supervision). His letter was full of hope. "The eyes of many are upon me; the newspapers have something to say about the matter often; and even the playhouse has its strokes of wit on occasion. The following lines begin an epilogue spoken the other evening by the celebrated Mrs. Jordon:

Cunning projectors may pretend to find
A scheme for sailing ships against the wind;
But never poet yet could start a scheme
For navigating plays against the stream."

All of this attention was making Rumsey a little uneasy, however. He continued, "This may truly be called the crisis of my life, should I succeed, I shall gain more reputation than I ever thought possible to fall to the share of any one man; if I fail, I shall be ridiculed and abused in all the public prints in Europe."[24]

Later that month Rumsey traveled to Paris, presumably to obtain a French patent.[25] It was there he met Jefferson for the first time. Jefferson was by now a fan of Rumsey's, impressed from all he had heard about the inventor from Madison, Washington, and Paine (who wrote to Jefferson from London around that time and said that Rumsey "appears to me perfectly master of the subject of steam, and is a very agreeable man").[26] Rumsey was equally impressed with Jefferson. He wrote to Morrow in Shepherdstown, "I have been frequently at Mr. Jefferson's, our American ambassador, he has got all that ease, affability, and goodness about him that distinguishes him as a good as well as a great man, he has taken much pains indeed to serve me."[27]

Jefferson arranged a meeting for Rumsey with Jean-Baptiste Le Roy, who was a member of the Royal Academy of Arts and Sciences. Le Roy told Rumsey that he already knew of him from letters he had recently received from his friend Dr. Franklin and that he was "much pleased with my plans." In his letter to Morrow, Rumsey described the proper dress for an afternoon meeting of this nature: "I was obliged to be dressed in a black coat, west coat, breeches, and stockings, my hair handsomely dressed and powdered, and the hind part in a large black bag; by my side a sword, my hat in my hand and (hard at my heels) a lusty French servant brought up the rear! . . . These things, Charles, that at first I had no idea was a necessary connection of a steamboat!"[28]

While in Paris that spring, Rumsey also met the American poet and businessman Joel Barlow, whom he described as a "steady, clever man." Barlow was well-known in America for his epic poem "Vision of Columbus," published in 1787. A Yale graduate, he had sailed to France in 1788 to serve as the European sales agent for the Scioto Associates.[29] This was a land-speculation company set up by a number of well-connected Americans who arranged, in secret with Congress, to buy from the federal government several million acres in the Ohio Valley at a cost of about ten cents an acre. They would pay the government for the land only after they had resold it at huge profits to European investors looking to send colonies to America. Barlow instead began selling small plots to individuals. He had no trouble finding buyers; French families were eager to escape the hardships of the French Revolution, and they were easily swayed by a sales brochure filled with exaggerated claims, the product of Barlow's British partner William Playfair. The two collected money from nearly four hundred subscribers, but Barlow mismanaged the funds and the Scioto Associates never paid the government for the land.[30] In 1790, several hundred French immigrants arrived in America, only to be told they owned no land in Ohio. A recent history of the scandal points to Barlow as "the primary culprit, whose blend of self-indulgence, incompetence, and perfidy was quite remarkable."[31]

At the time, Rumsey was considering using Barlow as his agent to obtain patents in several European countries, an idea Jefferson had encouraged him to pursue. Rumsey wrote to his American artist friend, George West, in London to ask him to discover "Mr. Trumble's [most likely the American artist John Trumbull, who was then living in London] opinion of [Barlow]."[32] There is no record of Trumbull's response, but Rumsey never hired Barlow, engaging him only briefly as an intermediary for letters he sent to Jefferson.

Rumsey returned to London in early April to oversee the building of his steam engine. He wrote to Jefferson in May that progress was slow. "The machine for my vessel has not gone so briskly as I expected, the case I believe with all new inventions the mechanics not being able to execute them with such dispatch as they do those they are acquainted with." He estimated that it would be ready in a month or so.[33]

In June, with time on his hands, he again wrote to Jefferson saying that he had heard about the new federal government taking form ("his Excellency General Washington arrived in Philadelphia on the 20th of April, amidst the acclamations of a joyous multitude, that the next day he set out for New York, . . . and that Congress was opened with great harmony. . . ."). The time seemed right to push Jefferson to consider a topic dear to his heart: the rights of inventors in the new national government. Rumsey told Jefferson that he had heard a bill was being considered in Congress that would set up an office for granting exclusive rights to inventors.

> . . . The dispute between Mr. Fitch and myself has caused many of the gentlemen to be very tenacious about giving grants, so much that the assembly of New York, and some others, would not give me a grant for the principles

of my boiler, but only for one formed like the drawing laid before them (which was intended only to explain its principle more clearly than expressions could), alleging that any other kind of grant would cut off others from improving on it, and so I think ought for a limited time, or what a grant would be worth, if every form that a machine can be put into should entitle a different person to use the same principle . . . where the principle itself is new I humbly conceive that it ought to be secured to the inventor for a limited time, otherwise but few persons will spend their money and time making new discoveries. . . .

I have troubled you, sir, with these remarks not only because I am deeply interested myself in having a just and permanent establishment of this business made, but because I wish my countrymen to have such encouragement given to them as to cause them to outstrip the world in arts and sciences.[34]

While Rumsey waited for his engine to be completed, his boat had been finished and was making its way from Dover to London. To his dismay, he found out that the man who had offered to pay for its construction, Whiting, had been sent to debtors prison before putting down a penny. Fortunately, Rumsey had enough money from the Rumseian Society to make the first payment. He then had to borrow money wherever he could find it— £500 from "a good-natured idiot," £200 from two other men, and the rest, including at least one thousand guineas needed to pay for his engine, in credit from his suppliers.

He felt the Rumseian Society had deserted him, but when he read the fine print he realized that it intended to establish credit for his work in London only *after* he had a successful run of his steamboat. He wrote of this problem to Morrow, saying, "I have written them [the society] a spirited letter in which I told them as this experiment would be at my own risk, that I expected all the profit if any should arise of that vessel." He added that he had named his new boat the *Columbian Maid* but noted, not at all modestly, that he planned to change it to *The Rumseian Experiment* "as soon as success is ascertained."[35]

But success seemed always to be around the corner. Problems with backers continued to plague him. In a progress report to Jefferson in September, he described how one of his major investors backed out on short notice, demanding that Rumsey pay him back within four days' time or he would sell the boat. (Because Rumsey was an American citizen, he couldn't register the vessel himself; the investor had to put it in his name.) Fortunately, Rumsey recalled that the man had promised in writing to pay for the engine. Turning the tables, Rumsey asked the iron founder who had built the engine to show a copy of their agreement to the man and demand immediate payment. "This frightened him so that he came to terms immediately," Rumsey told Jefferson.[36]

Rumsey finally had no choice but to go back to his agents in London, Vaughan and Barclay. They lent him enough money to pay all his debts. But, he wrote Jefferson, "this unlucky affair has put me back near a month with my experiment." He was still hopeful, though. He described a successful test of the engine at the dock, and from that he predicted that his boat would easily make ten miles an hour.[37]

That same week, Jefferson wrote to Rumsey with the news that "there was a difficulty with the committee of the [French] academy, arising as [Le Roy] apprehended from their not understanding your principles." Jefferson told Rumsey that he and Le Roy had decided not to "endanger final success by prematurely pressing the Academy" and instead would wait until Rumsey had proved his steamboat in London.[38]

Two weeks later Rumsey wrote again to Jefferson (who in the meantime had sent another letter to Rumsey inquiring of his progress) that he was still facing delays caused by workmen and hoped to try the boat in October.[39] Another letter to Jefferson dated October 4 asks for his patience. Rumsey again blames delays in readying his steamboat for its first trial.

Jefferson left France that October to return to the United States; Washington had asked him to serve as secretary of state in the new government. Before he left, he wrote once more to Rumsey and asked to be notified when his first trial was made. He assured Rumsey that as soon as the steamboat was a proven success, "Mr. Short [Jefferson's secretary in Paris, who remained there in a diplomatic position] will do for you in Paris whatever I could have done in obtaining you a patent there."[40] Six months passed with no news. Jefferson wrote to Short in Paris in April 1790. "What has become of Rumsey and his steamship? Not a word is known here. I fear therefore he has failed."[41]

Jefferson guessed correctly. Sometime during the fall or winter of 1789, Rumsey gave the *Columbian Maid* her first trial. No details were ever recorded as to what happened, but Rumsey was so depressed that it wasn't until February that he wrote to anyone about it. In a letter to Morrow he said, " . . . every possible disappointment attended my experiment, when it was all put together it proved so imperfect that almost the whole of it has to do over again, the great delay and enormous expense attending it, made my friends doubtful and uneasy; and thereby put it out of my power to obtain money from them, to pay my bills, which daily come upon me."[42]

He then described how he had spent two weeks hiding from the London bailiffs, who were looking to throw him into debtors prison. In the nick of time he was able to borrow money from a few new friends (including the "son of Parson West") to pay his suppliers. But his problems didn't end there. When his workmen heard of his financial difficulties, they "came on me at once," demanding to be paid immediately. At this point, in desperation, Rumsey resorted to a bit of psychological warfare. He set a time and a place to meet with the workers, most of whom he had never met. In a letter to a friend, he recalled how he dressed himself " . . . in style, I therefore had my head as large and as white as a lord's wig, I told them that if I paid them they would instantly quit my employ, as I could get a number that would be glad of my work and would wait as *usual* until six months after the work was done for their pay, this maneuver and an air of importance that I forced on (much against my natural inclination) made the mean rascals (for such the most of them are) bow to the ground and tell me with one consent that they would finish the work, if I would let them have that *honour.*"[43]

With his workers pacified, Rumsey went back to work. But as soon as he began ordering parts, he was hit with another setback: His suppliers had cut off his credit. By March 1790, Rumsey needed cash badly. Fortunately, he had become friendly with two American businessmen living in London, Daniel Parker and Samuel Rogers. The men seemed intrigued by the idea of building a steamboat.

SIX

LORD HIGH ADMIRALS
OF THE DELAWARE

Progress is not an illusion, it happens, but it is slow and invariably disappointing.

—*George Orwell*

Back in Philadelphia in the early months of 1790, the disappointing steamboat experiments of the previous fall bore down on John Fitch. Not only was he broke and dependent on Mary Krafft for a place to live, he was being browbeaten by his Steamboat Company shareholders. "Dr. Thornton and Mr. Stockton have treated me with such indignities as a man in my station ought to resent, whereas had I been on equal footing with them, I probably should only [have] laughed at them." He added that they "also charged me with stupidity, botching, being a man [who] could not be depended upon, with drinking, tippling, and every opprobrious name that could be artfully invented."[1]

By early March, when he began getting the boat ready for a fresh season of experiments, Fitch's ability to take criticism had reached the breaking point. "When in easy circumstances, [I am] modest to excess, and put up with almost any indignities, and resent them no other way than by a familiar levity, but when in wretchedness [I am] haughty, imperious, insolent to my superiors, tending to petulance. . . . And a man of this disposition . . . can never get through the world easy."[2]

Over the winter, the Steamboat Company directors agreed that Voigt's pipe boiler, with its frequent leaks, was causing more problems than it was worth. Against Fitch's wishes, they ordered a traditional pot boiler built to replace it. But that wasn't all. The condensers that Hall and Thornton had developed the previous fall proved no better than anything tried before. Now Thornton wanted to try again, this time with a condenser that was twice as large as his last one. Again, Fitch opposed the idea, but the directors ignored him and gave Thornton the go-ahead. A trial was set for Easter Monday, in early April. As Fitch watched from the shore, Thornton and Stockton took the boat out on the Delaware only to find that the new machinery barely moved the boat against the current. Bigger obviously was not better.

Fitch walked back to his room in Mary's tavern to try to figure out where the problem was. He analyzed every part and every mechanism and finally decided the weakness had to be in the condenser. He recalled that the small, straight pipe condenser in the 1787 boat had been the most powerful of all, relatively speaking, because its design provided a more perfect vacuum than anything they had tried since. He took up his pen and began to lay out his thoughts: "Thornton's condenser is undoubtedly one of the best calculated to condense without a jet of water, but I conceive the difficulty of getting rid of the air is insurmountable. . . . [But] when [the air] is drove back again by the steam to the cold condenser, it becomes nearly equal to common air in density, and skulks into the bottom of the condenser for security. Here it cannot be dislodged until the steam is destroyed."[3]

Fitch had tried adding a separate air pump to remove the excess air back in 1787, but that alone didn't solve the problem. Now he had a new idea: He proposed shaping a small condenser in the form of a straight tube, in which "the quantity of air remaining would be inconsiderable to what would be in a large condenser, consequently less capable of injuring us and a much more perfect vacuum formed." Fitch laid out his plan to Thornton and a few of the other directors, and they agreed to give it a try.

Within a week, Fitch and Voigt had made the necessary alterations to Thornton's condenser. On a Monday in mid-April, as Thornton and Richard Wells watched from the shore near the Front Street wharf, "We went out in the river and found our engine to work very favorable, so that in a short time it [broke] one of our pulleys in two that the main chains acted upon, which obliged us to come to anchor in the middle of the river. . . . We lay at anchor several hours, and in the meantime came by five or six boats, and every one refused assistance, and most of them exulted and seemed to have a heartfelt pleasure."[4]

Even as they sat helplessly afloat in the middle of the river, the two inventors must have been ecstatic about the engine's newfound power. Four days later, after reinforcing the pulleys, they tried again. This time, "altho the wind blew very fresh at N.E., we reigned lord high admirals of the Delaware, and no boat on the river could hold way with us, but all fell a-stern." Fitch then gleefully described how several sailboats tried—and failed—to race past them. "And [we] came in without any of our works failing. Which fully convinced us of what we had pursued so long, and with such embarrassments. We being flushed with success, and knowing that 4 men could navigate 100 tons up the Mississippi, we concluded that our troubles were at an end, and agreed to invite the Company to see her perform."

Eager to show off their new engine, they invited the directors for a ride the next day at two o'clock. But "at the time appointed blew a perfect storm, and none of the members came except Mr. Say." A horrendous wind had whipped up, and Fitch tried to talk Voigt into postponing the demonstration. But Voigt insisted they would be safe, so the three men took a quick trip and returned without incident.

A few days later they "took a sail about four miles and back" with scientists Robert Patterson and David Rittenhouse, who had once called the steamboat idea "ridiculous," and a few other influential Philadelphians, hoping that they would spread the news of the boat's success around town. Fitch by now was certain of the boat's potential to navigate the Mississippi, and not a little proud: "This will be of the first consequence to the United States, and make our Western territories four times as valuable as otherwise it would be. This has been effected by little Johney Fitch and Harry Voigt, one of the greatest and most useful arts that was ever introduced into the world."

At last the steamboat was deemed worthy of a positive mention in the local press. In early May, the *Gazette of the United States* printed this report from Burlington, New Jersey: "The friends of science and the liberal arts will be gratified in hearing that we were favored, on Sunday last, with a visit from the ingenious Mr. Fitch, accompanied by several gentlemen of taste and knowledge in mechanics, in a steamboat constructed on an improved plan. From these gentlemen we learn that that they came from Philadelphia in three hours and a quarter, with a head wind, the tide in their favour. On their return, by accurate observations, they proceeded down the river at the rate of upwards of seven miles an hour."[5]

The next step was to get the boat ready for commercial service. With the profitable summer months almost upon them, the shareholders didn't want to waste any time in recovering their investment. Fitch had spent nearly $4,000 on the project since he began in the spring of 1786, an enormous amount in those days. The next expenditure was to make the boat comfortable for passengers.

Thornton, the budding architect, showed the directors and Fitch his sketches for an elaborate passenger cabin. "A pretty considerable dispute took place between Dr. Thornton and me, respecting the height." Fitch was afraid that a tall cabin would create wind resistance and slow the boat's speed. "And knowing him to be a hasty, opinionated man, [I] gave my objections to him in writing." Fitch wryly suggested to Thornton, " . . . if it must be elegant, make it low and line it with gold!"[6] But Fitch lost another battle to Thornton, who never failed to remind Fitch that he was the majority shareholder. At this point in the venture, Thornton reasoned, appearance and passenger comfort were more important than speed.

By mid-June, the cabin was completed. The Steamboat Company decided to inaugurate service by inviting the governor of Pennsylvania, Thomas Mifflin, and the state executive council for a ride. They accepted. After their successful journey, the governor told Fitch to order a set of flags to fly on the boat at the council's expense. Fitch did so, and when the flags arrived he tried to meet with Mifflin to arrange a date for some sort of presentation ceremony. Each time, the governor refused to see him. Fitch went to the secretary of the council, a Mr. Biddle, who promised to look into the matter. Three months later, Biddle told Fitch that the gift of flags was a personal offering by the council members, not an official state purchase; a public ceremony was not going to take place. Evidently, the politicians feared granting official recognition to a project that many still viewed as a crazy venture.

With the boat's success that summer, the time had come for an important test. More than a year earlier, Fitch had formed a second company of shareholders after the almost complete dissolution of the first. Under its terms, once the boat moved at eight miles per hour, the second company's forty shares would be turned into the original company and all shares would become of equal value. As Thornton recalled the event twenty years later,

> . . . a mile was measured in Front Street (or Water Street) . . . and the bounds projected at right angles, as exactly as could be to the wharves, where a flag was placed at each end, and also a stop watch. The boat was ordered under way at dead water, or when the tide was found to be without movement; as the boat passed one flag it was struck, and at the same instant the watches were set off; as the boat reached the other flag, it was also struck, and the watches instantly stopped. Every precaution was taken before witnesses; the time was shown to all, the experiment declared to be fairly made, and the boat was found to go at the rate of eight miles an hour, or one mile within an eighth of an hour; on which the shares were signed over with great satisfaction by the rest of the Company.[7]

On June 15, 1790, the *Pennsylvania Packet* newspaper carried this historic first advertisement: "The Steamboat is now ready to take passengers, and is intended to set off from Arch Street, in Philadelphia, every Monday, Wednesday, and Friday for Burlington, Bristol, Bordenton [*sic*], and Trenton, to return on Tuesdays, Thursdays, and Saturdays. Price for passengers, 2/6 to Burlington and Bristol, 3/9 to Bordentown, 5s. to Trenton." Fitch placed twenty-three ads that summer in the *Packet* and in the *Federal Gazette*.[8]

Estimates of total miles run that summer range from 1,300 to 3,000. According to Fitch, the steamboat often went for 500 miles between breakdowns—a remarkable record, especially considering that several "accidents" were caused by Voigt's habit of hanging a weight on the safety valve to get more power. Fitch notes that most of the repairs were minor and could be made in less than two hours. *New York Magazine* published a correspondent's report dated August 13, 1790, that began, "Fitch's steamboat really performs to a charm. It is a pleasure, while one is on board her in a contrary mind, to observe her superiority over the river shallops, sloops, and ships etc., who, to gain any thing, must make a zigzag course, while this, our new intended vessel, proceeds in a straight line. . . . Fitch is certainly one of the most ingenious creatures alive, and will certainly make his fortune."[9]

But by then Fitch realized that his fortune wasn't going to be made anytime soon. On every trip, the steamboat lost money. Travelers were slow to come around to the idea of riding on a wooden boat with a fire blazing on board. They preferred stagecoaches, even though the steamboat was faster and half the price. Desperate to attract customers, the Steamboat Company directors told Fitch to offer free beer, rum, and sausages in the boat's elegant little cabin, but even those enticements didn't work. On Saturdays, they tried running a special service to Gray's Gardens, a popular summer drinking spot and dancehall on the Schuylkill River, undercutting the price offered by a regular ferry service. They still couldn't find enough passengers.

Fitch remained convinced that the real value of his boat was on the Ohio and Mississippi Rivers. There, without stagecoach competition and with the potential for upriver trade, which was then virtually nonexistent, the possibilities seemed limitless. Fitch's Virginia monopoly specified that he had to be running two steamboats on the state's waters by November 7, 1790, just two months away. There was little time to waste. Losing this monopoly would mean losing the rights to navigate on the Mississippi, since in those days Virginia's borders extended to its eastern banks. In August, the Steamboat Company approved plans to start building a sturdier boat—presumably to be able to take it out to sea in order to reach Virginia—to be called *Perseverance II.* The company assessed each shareholder £10, but to Fitch's great dismay it didn't require upfront payment. This meant he would have to build the boat by pleading with his suppliers for credit, something he hated to do.

In an attempt to get government aid, Fitch paid visits to several Pennsylvania Assembly delegates from the western part of the state, including General John Gibson and Albert Gallatin (a financier and secretary of the treasury under Jefferson and Madison who, ironically, would publish in 1808 a government master plan for improving the transportation infrastructure in the United States). Fitch tried to convince these men that his plan to build steamboats in Pittsburgh would greatly benefit their constituents. For the past several years, farmers and merchants in the state's western counties had been struggling. The cost of shipping goods across the Alleghenies to the eastern cities was more expensive than the cost of the goods themselves, which led to very high prices. In addition, since 1784 they had not been able to transport their products down the Mississippi to New Orleans. In that year, the Spanish, who owned the territory west of the Mississippi and controlled New Orleans, had closed both the river and the port to all but Spanish goods. Except for Gibson, who toyed with Fitch for a while before rejecting him, the other western Pennsylvania representatives refused to even listen.

Fitch tried another approach. He sent a detailed letter to the well-known Philadelphia financier and then U.S. Senator Robert Morris proposing that Morris establish a trading house in New Orleans with his former colleague Oliver Pollock, a successful merchant in New Orleans for many years.[10] Fitch offered a deal that included giving the partners the right to build steamboats for use on the Mississippi and sharing the profits with other investors, including Fitch and Voigt. Fitch even declared he would move to New Orleans and become a Spanish citizen, something that must have shocked Morris and Pollock, since both men were fervent patriots who had lent financial support to the Revolution and other U.S. causes in the past. Morris sent back a curt reply, saying that before they could do anything, Fitch would need to obtain a permit from the Spanish governor of Louisiana.

Between appeals for money, Fitch continued to help Voigt and their workers to get *Perseverance II* ready to meet the November deadline. Although they had hoped it would be relatively easy to duplicate the workings of their existing boat, Fitch and Voigt soon became frustrated. Every part was one of a kind, original designs made by different workers at different times. Standardization of parts, a concept put into practice by the inventor Eli Whitney several years later, did not yet exist. As the days passed, short-cuts were taken and substitutions were made, just to put together some-thing—anything—that could move upstream in Virginia waters.

As Fitch's luck would have it, on the night of October 13, as the boat was nearing completion, a nasty storm blew through Philadelphia. Winds ripped the boat from her moorings and blew her across the river to Petty's Island. It took more than a week to make repairs and move the boat back to shore. The Virginia deadline requiring Fitch to have two steamboats in the state's waters was now two weeks away. Fitch wrote a desperate letter to the Vir-ginia Assembly pleading for an extension. In early November, a committee reviewed his request and for reasons unknown deferred action on it until March 31 of the following year. Fitch tied up his two boats for the winter, not sure if they would ever run again. Now his last hope was to receive a U.S. patent, which he had applied for that fall; the Patent Act had been enacted several months earlier. Receiving the first patent for steamboats would settle once and for all the dispute between him and Rumsey. More important, it would attract a fresh lot of investors.

By this time, he had sold his last share of company stock and was poorer than he'd ever been. Speaking of himself, he wrote, "from an overzealous zeal to render real service to his country he has reduced himself from a state of easy liv-ing to a state of penury."[11] Desperate for a steady income, he began to apply for the government positions that were rapidly opening up now that the capital had moved to Philadelphia. His first choice was a position with the future U.S. Mint. In the winter of 1790, with the new government just recently opened for business, Fitch wrote a letter to his friend, the politician and educator William Samuel Johnson, asking for advice on the best way to approach President Wash-ington for mint appointments for him and Voigt. He ended the letter with a postscript: "If there is but one appointment let it be Mr. Voigt as a person more capable than myself."[12] But Congress had not yet established the federal mint.

Fitch soon realized that a mint job, if one ever materialized, was too far down the road to help. In early March he wrote to the secretary of war, Henry Knox, asking to be considered for appointment in the western army in a position involving gun smithing or surveying, citing his earlier experiences in both fields.[13] He also applied for the newly created position of job of sergeant of arms at the Pennsylvania Assembly but was quickly rejected. That same month, he ran an ad in one of the local newspapers with the heading, "John Fitch, Clockmaker and Goldsmith, begs to inform the public that he proposes carrying on his trade, in Second Street, next door to the Sign of General Muhlenburgh, in Campington."[14]

In utter desperation, on Christmas Day 1790, Fitch wrote an impassioned letter to the few remaining members of the Steamboat Company. He reminded them of their efforts and successes and asked their permission to make a proposal to Congress: For a grant of 50,000 acres of western land, he would take a steamboat from the mouth of the Mississippi to the Ohio rapids. With no more company shares left to sell, he asked the directors to support him for three more months.[15] Apparently they did, although their response to his plea is unknown.

Money problems aside, Fitch also suffered a personal setback that winter. The threesome of Fitch, Voigt, and Mary Krafft had reunited earlier in 1790, and Fitch felt sure the affair between his two friends had ended. But he had been blind; in fact, their illicit relationship "continued till the effects of love promised to further increase to their families."

> . . . They both being ashamed to let me know their offense she went from home four or five months and happened to fall in with some of her acquaintance and you may well think the predicament she was thrown into. Not knowing how to protect her honor and thinking I was an easy disposition and her sincere friend and had made such a generous offer about two years before, [she] ran the venture without my leave of throwing herself on me for protection and *called herself after my name.* [Emphasis added] This when I heard not only alarmed [me] but I thought it unkind usage and the first time when I knew that one person could make a bargain. I forgive her. And had the child been so much of my complexion as that I could have fathered it with repute it would have given me a very little uneasiness as knowing her to be a valuable woman and one ever faithful to the man she loved.[16]

He then described the events surrounding the birth of Mary's second child with Voigt. "She was brought to bed on her return home about eight miles off and word sent to me as her husband." This request disturbed Fitch no end. "But on the other hand I knew the goodness of her heart and that she was led into her errors by the purest love. . . . Considering her deplorable situation I did then forgive her and do still forgive her and hope that the purest friendship may ever subsist between us."[17] Although Fitch may have gone to see her at that time, it seems he turned a deaf ear later, as evidenced by a letter she wrote to him pleading for him to visit her as he had promised; she signed the letter, "your loving wife, Mary Fitch."[18]

She returned with the baby to Philadelphia, and to Fitch, who was still living at her tavern. The neighbors presumed the child was his. He wanted to marry her, but he was determined to wait until her ardor for Voigt had completely cooled. Fitch seemed to have forgotten that he was still married to Lucy, the woman he had left nearly twenty years earlier. Fitch's son, Shaler, was by now a young man. Fitch had never seen his daughter, also named Lucy, who was born after his departure. Desertion was more common than divorce in those days, and it was also the most frequently cited grounds for the divorces that were granted. It is not known whether Lucy ever divorced Fitch, but it seems she never remarried.

Fitch found he could not extricate himself from this sad situation. "I believe that the greatest torment that a man can have in this world is to be teased with a woman. And have ever been of that opinion since I left New England. And not being a very handsome man and one of very indifferent address and of no flattery. . . ." He stayed with Mary, worried not only about her reputation but his. If he left her now, he reasoned, he would be looked on as a "damned rascal." In fact, Fitch was in love, despite having sworn off romantic entanglements some twenty years earlier. "I esteem her more virtuous than nine-tenths of women."[19]

Recounting these events hit him hard as he wrote about them several months later. Certainly he had harbored hopes that her love for Voigt would cool and her love for him would grow. Neither happened. "All I can say of the matter is this, I think this is a damned wicked world and when I get clear of it [I] never wish to come back to it any more. I have frequently been apt to conclude that it is a place where they transport souls to from other planets that is not fit to live in them, the same as Great Britain used to send convicts to Virginia. And if I was sent here as a lunatic to bedlam . . . I am sure [I will be] more cautious when I get back to Jupiter again."[20]

━ ▬ ━ ▬ ━ ▬ ━ ▬ ━ ▬ ━ ▬ ━

Around this time, while waiting for his hearing with the U.S. patent commissioners, Fitch was offered a chance to build a steamboat in Europe. Aaron Vail, the U.S. consul in the French port city of Lorient, was visiting Philadelphia that winter and had seen Fitch's steamboat. The two men met, and Vail was sufficiently impressed to offer Fitch a deal. They signed a contract on March 16 in which Vail would procure a patent from the French government in Fitch's name, as well as pay for at least one steamboat to be built in France, including the costs of a mechanic. In return, Vail and Fitch would share half the profits from any French steamboats they built, and Fitch in turn would reimburse the members of the Steamboat Company for their entire investment.[21] A year later, Vail would send word to Fitch that the French government had granted his patent.

SEVEN

The First and True Inventor

The fact is, that one new idea leads to another, that to a third, and so on through a course of time until someone, with whom no one of these ideas was original, combines all together, and produces what is justly called a new invention.

—Thomas Jefferson

By early 1791, after months of delay, the deciding battle of the steamboat wars was about to be won. For nearly six years, Fitch and Rumsey had been vying for state monopolies and the right to be called the inventor of the steamboat. The grand prize of the first federal patent for steamboats was about to end the fight. A hearing for the steam engine and steamboat patent applicants before the newly established Patent Board was set for the first Monday in February.

The nation's first patent act had been signed into law in April 1790, the result of a last-minute clause inserted into the Constitution three years earlier. The delegates to the Constitutional Convention spent the summer of 1787 debating many weighty issues, but intellectual property was not one of them. The rights—and potential profits—of the few who wanted to protect their creations ranked low in the grand scheme of things. Patents for "useful arts" and copyright protection for written works were considered to be two sides of the same coin in those early days.

Prior to federal patent law, state legislatures tried to encourage inventors by offering a period of exclusive use for their creations, the terms and times of which varied from state to state. Except for South Carolina, which tacked a patent clause onto its 1784 copyright law, patents as we know them today did not exist. The need to petition states one at a time put a huge burden on inventors, and few had the time or money to follow through. According to one count, only twenty-three state "patents" were issued between 1779 and 1791.[1]

Copyright protection had come a little further by 1787, but not by much. Prior to the first federal copyright law, which also was enacted in 1790, authors who wanted to protect their writings from piracy faced similar challenges. Before independence, English authors whose publications were sold in America were covered by English copyright law, which had been around since the Statute of Anne in 1710.[2] American authors were mostly out of luck. After the Revolution, several writers began asking state legislatures for copyright protection for their books. One was a young American lawyer and teacher named Noah Webster, who would later publish a dictionary that would shape the way Americans spoke and spelled. Details are sketchy, but it appears that on separate occasions he and his Yale roommate, Joel Barlow, spent time in New York in the early 1780s urging members of the Continental Congress to enact some sort of nationwide copyright law.[3]

Under the weak Articles of Confederation, however, Congress had no authority to enact this kind of federal legislation. The best it could do was to name a three-man committee, which included James Madison, to make recommendations to the states on the matter.[4] In May 1783, the committee issued a resolution urging each state legislature to pass laws providing copyright protection for a term of not less than fourteen years to U.S. authors and publishers.

In the meantime, Webster returned home to Connecticut to lobby his own state legislature to pass a copyright law, the first of the states to do so. Over the next few years, while traveling across the country to push sales of his textbooks, he urged other state legislatures to pass similar laws. His efforts surely helped; within three years, twelve of the thirteen states (all but Delaware) had a copyright law on the books.[5] One of those states was Virginia, where Madison had returned to serve in the assembly. During a trip through the south, Webster stopped in Richmond and persuaded Madison to sponsor Virginia's copyright bill, which was passed in the fall of 1785. Interestingly, this was around the time that John Fitch was in Richmond convincing Madison to present his petition for a Virginia steamboat monopoly.

Perhaps, then, it was no coincidence that Madison was one of two delegates who proposed adding these words to the U.S. Constitution on August 18, 1787: "to secure to literary authors their copyrights for a limited time" and "to encourage by premiums and provisions, the advancement of useful knowledge and discoveries." The other was Charles Pinckney of South Carolina, who on that same day asked that clauses be added to "grant patents for useful inventions" and "to secure to authors exclusive rights for a certain time."[6]

At least two events that historic summer in Philadelphia may have alerted the delegates to the need for such a clause. The first occurred on August 7, when the inaugural meeting of the Pennsylvania Society for the Encouragement of Manufactures and the Useful Arts took place. Tench Coxe, a Philadelphia merchant and outspoken advocate of American manufacturing and industrial development (who a few years later would serve as assistant secretary of the treasury under Alexander Hamilton), spoke to the group and encouraged the government to take an active role in encouraging innovation: " . . . Premiums for useful inventions and improvements, whether foreign or American . . . must have an excellent effect. They would assist the efforts of industry, and hold out the noble incentive of honourable distinction to merit and genius. The state might with great convenience enable an enlightened society, established for the purpose, to offer liberal rewards in land for a number of objects of this nature."[7]

The second event was the successful running of Fitch's steamboat *Perseverance* that summer, which was the talk of Philadelphia. On August 22, just four days after Madison and Pinckney proposed adding copyright and patent clauses, Fitch seized an opportunity to influence the contents of the Constitution. He invited William Samuel Johnson, who was then a convention delegate from Fitch's home state of Connecticut, for a ride on his steamboat and encouraged him to bring along other interested delegates, which Johnson did. As Fitch wrote later, "There was very few of the convention but called to see it, and [I] do not know whether I may except any but General Washington himself. . . . Governor Randolph with several if not all of the Virginia members were pleased to give it any countenance they could."[8] Certainly Fitch would not have been shy about pushing his guests to support a federal patent law. The day after the demonstration, Johnson sent a note to Fitch offering his "compliments" and the promise of encouragement in the future.[9]

Several days later, a member of the Committee of Details, probably Madison, combined the wording for copyright and patent protection into a single clause. It was approved, along with several other additions, on September 5, without debate or a dissenting vote. Placed in Article 1, Section 8, it reads, "The Congress shall have the Power . . . to promote the Progress of Science and useful Arts, by securing for limited Times to Authors and Inventors the exclusive Right to their respective Writings and Discoveries." Madison's original wording, which called for "premiums and provisions," was later eliminated because of worries about future financial obligations to the federal government. On September 17, the Constitution was approved. When it was ratified in 1788, state patents and copyrights became superfluous, although New York and a few other states continued to award them for many years to come.

Patent law had been around for several centuries, beginning in Venice and Germany in the 1400s. Then, as now, there were requirements for models and examination by a board, and a term of ten to twenty years was typical. By the eighteenth century, the concept of intellectual property—the idea that a person can own his original ideas just as he can own material property—had spread to France, England, and other parts of Europe.[10]

Not everyone was in favor of patents, most notably Thomas Jefferson. He long believed that "the exclusive right to invention [is] given not of natural right, but for the benefit of society."[11] To prove his point, he never patented any of his several inventions, which included a moldboard plow, a wheel cipher for decoding secret messages, and a portable copying press. (Only one

U.S. president has ever received a patent: Abraham Lincoln, in 1849, for his device for buoying vessels over shoals.) As one historian put it, Jefferson "conceived government's role to be that of disseminating information useful to citizens rather than insuring material profit for inventors."[12] It is interesting to speculate on what might have become of the patent and copyright clause if Jefferson had been a delegate to the Constitutional Convention rather than minister to France in 1787.

Jefferson had written to Madison from Paris in July 1788 to argue for adding a bill of rights to the Constitution that would, among other things, forbid monopolies, including the term monopolies inherent in patents, arguing that "the benefit even of limited monopolies is too doubtful to be opposed to that of their general suppression."[13] Madison disagreed, replying, "With regard to monopolies they are justly classed among the greatest nuisances in government. But is it clear that as encouragements to literary works and ingenious discoveries, they are not too valuable to be wholly renounced?"[14]

The patent and copyright issue was one of the first items the new government took up. On April 1, 1789, enough members had arrived in New York to form a quorum, and the First Congress set to work. After being pestered almost daily by authors and inventors seeking government protection for their creations, the legislators figured they had better put a system in place for awarding patents and copyrights, if only to allow them to get their other work done. On April 20, a bill was ordered to "be brought in, making a general provision for securing to authors and inventors the exclusive right of their respective writings and discoveries."[15] Although no draft documents of this bill have survived, Noah Webster may have written at least one early version of this bill, according to a brief entry he made in his diary in early April.[16]

Once the news reached Fitch and Rumsey that a patent law bill was in the works, they stepped up efforts to influence the contents. Fitch wrote to his old supporter, now a U.S senator, William Samuel Johnson, and asked for details of the bill. To play it safe, on April 12, Fitch submitted yet another petition (his third since 1785) for steamboat rights; it was read on May 13.[17] May was a busy month for patent applicants. On May 4, one Alexander Lewis petitioned the House for a patent for propelling boats (his plans remain unknown; later patent office records show no reference to him); on May 6, Arthur Green asked for a patent for his device to measure longitude; and on May 14, Englehart Cruse asked for a patent for a steam engine to be used in mills.[18] In all, Congress would receive twenty-four petitions for patents before the patent act was signed into law.[19] All were tabled until a law could be passed.

Rumsey, still in London, had expressed his thoughts on patent law to Jefferson in a letter a year earlier. His recent experience with British patent law led him to believe that it didn't go far enough in encouraging ingenious men, "which I suppose was the object of such laws was intended to embrace." He told Jefferson he supported the process of examination, as the French did, saying that the practice "has a tendency to prevent many simple projectors from ruining themselves by the too long pursuit of projects that they know but little about."[20] Examination involves verifying, among other things, that the invention works as claimed and that it has not been previously invented.

On June 23, House Bill 10, calling for a combined copyright and patent act, was introduced. But the representatives sat on the bill all summer and into the fall. When President Washington delivered his state of the union address to the reconvened Congress on January 8, he made it clear that he expected the legislators to quit stalling and move quickly to pass a patent bill, and to make sure it included language that allowed for inventions from other countries to be patented in the United States: "The advancement of agriculture, commerce, and manufactures, by all proper means, will not, I trust, need recommendation; but I cannot forbear intimating to you the expediency of giving effectual encouragement, as well to the introduction of new and useful inventions from abroad, as to the exertions of skill and genius in producing them at home."[21]

In early February the House separated the single bill into two, one for copyrights and one for patents. A week later, two more steam engine inventors petitioned Congress for patent rights: Nathan Read of Massachusetts and John Stevens of New Jersey. The debate on the patent act went on until March 10, when the House passed a version to be sent to the Senate.

Fitch somehow kept himself informed on the contents of the draft bill and sent in a proposed change, which was read to the Senate on March 22.

Anticipating problems with Rumsey down the road, he asked that the act re-instate a clause in an earlier version of the bill that called for a trial by jury in case of disputes. He also asked for a clause to be added that would pro-tect state patents already granted, an issue Congress had so far avoided. Both his suggestions were ignored, and the patent act was signed into law by Pres-ident Washington on April 10. It was the tenth act passed by the First Con-gress, right up there with laws on duties and tariffs and legislation establishing the offices of the secretary of state, the secretary of war, and the attorney general.

The Patent Act of 1790 called for a board consisting of the secretary of state, the secretary of war, and the attorney general to examine and issue patents, which would be granted for fourteen years to any applicant that "hath . . . invented or discovered any useful art, manufacture, . . . or device, or any improvement therein not before known or used." It also required that the patentee submit a model of the invention detailed enough to allow an-other person to copy it once the patent term expired. The law reflected Fitch's strong beliefs in two key areas, perhaps a result of his influence: It dis-tinguished between improvements and original inventions, and it required that applicants publicly disclose their ideas. Secretive inventors like Rumsey would no longer hold an advantage over their competitors.

In addition, the new law, unlike those in England and France and against the clear wishes of the president, did not allow inventions already patented elsewhere in the world to be patented in the United States; U.S. patents had to be original, with no prior use anywhere. Apparently the legislators be-lieved that foreign technology would arrive eventually through the time-honored practice of piracy, as Samuel Slater demonstrated a short time later by bringing English textile mill technology to Rhode Island. The lawmakers wanted to encourage and give advantages to U.S. citizens only.

Within a week of the act's passage, the patent board opened for business at the Department of State offices in New York. Its lead commissioner, iron-ically enough, was the nation's first secretary of state, Thomas Jefferson, who had returned from France a few weeks earlier. The other two members were Secretary of War Henry Knox and Attorney General Edmund Randolph. They established a few basic rules, including one that instructed inventors to present their descriptions, models, and a small fee (less than $5) to Henry Remsen, Jr., Jefferson's chief clerk.

The steam patentees wasted no time in refiling their petitions. On April 16, according to the memorandum book at the Department of State, Nathan Read of Massachusetts filed for patents for "his inventions and improvements in the art of distillation, and in the cylinders and case [a reference to a portable boiler], made use of by Watt and Boulton." Read met with the patent board the following day, and on, April 22, the commissioners agreed to issue him a patent. The next day, Read filed another application, this one for a steamboat, a steam carriage, and a portable boiler.[22]

In mid-May, John Stevens applied for his patent, for "improvements respecting the generation of steam and the application thereof to different purposes," which included two types of boilers, an improvement to a Savery steam engine, and methods using pumps and pistons.[23] A few days later, Joseph Barnes, acting for James Rumsey, filed for patents for nine devices, including a steamboat, improved steam boilers, and several improvements to mills.

While he was at the State Department offices, Barnes must have been able to look at the records of previous applications, because on June 3 he protested the award of a patent to Read, claiming that Rumsey was the original inventor of the boiler Read proposed. The board agreed to delay issuing the patent until more information could be gathered. Were it not for Barnes's protest, Nathan Read would have been the first person to receive a U.S. patent, and for steam technology no less. If Barnes hadn't protested, the steamboat wars might have ended differently.[24]

On June 22, Fitch, who had to tear himself away from the business of running his steamboat service that summer, arrived in New York and submitted his application to "request a law in his favor, independent of the general one now in force" regarding "improvements made in applying steam to the purposes of propelling boats or vessels through the water."[25] The board met with him the next day and asked him to provide a model and more detailed specifications. Two weeks later Fitch filed another petition with the patent board, "praying that a patent may be refused to any other person than himself for propelling boats or vessels through the water by force of

steam . . . as he will prove that he was the first, true, and original inventor."
Like Barnes, Fitch must have also been able to read the previously filed applications. By now, the patent board could see problems ahead.

In November 1790, the federal government moved from New York to the temporary capital of Philadelphia, and the patent board began conducting business from the State Department office at Market and Seventh Streets. Later that same month, Fitch showed up to file another petition, this one asking that a rule be set requiring all claims, arguments, and proofs to be put in writing. He had seen Rumsey's supporters, including Barnes, in and out of the State Department offices and feared they were exercising undue influence on the commissioners, in particular Rumsey's friend Jefferson.

Fitch's petition must have spurred some sort of reaction among the patent board commissioners, because the following day Remsen sent a letter to Fitch and four other men who had applied for steam-engine-related patents: Rumsey, Stevens, Read, and Isaac Briggs. Remsen explained that the board did not feel the submitted applications were satisfactory and therefore wanted to "hear all those claims together." He told them to attend a hearing on the first Monday in February, in which all applicants for patents involving new applications of steam would be able to state their case.[26]

Fitch knew that he would be facing Rumsey in a federal patent fight, but it is not known when he became aware of the other steam-engine patent applicants. He would have recognized John Stevens as the person who supported his steamboat petition in New Jersey three years earlier but later dropped his offer of aid. In fact, Stevens backed out because he came to believe that Rumsey's ideas had more promise. He even wrote to Rumsey in 1788 offering him advice and ideas about pipe boilers.[27] Rumsey ignored him. Frustrated at Rumsey's lack of progress, Stevens petitioned the New York legislature for a steamboat monopoly of his own. But the state turned him down because of the law that was already in place for Fitch. Stevens was not discouraged, though, and continued to think about steamboats for the next few years. His unwillingness to get his hands dirty meant that his ideas were on paper only; he needed a mechanic to carry them out. Stevens was forty years old at that time, a third-generation American from a wealthy family. Educated at Kings College (later Columbia University), he was a colonel during the Revolution, at which time he invented—but never built—some sort of machine to destroy British warships.[28] After the war, he bought a large tract of seized English-owned land at auction that would become Hoboken, New Jersey, and built a home there.

Nathan Read was probably the most intriguing and capable of all the steam inventors. Little has been written about him, probably because so little came of his efforts. At the time of his patent application, Read was a forty-year-old Harvard graduate who owned an apothecary shop in Salem, Massachusetts. He became interested in steamboats as a result of the Fitch–Rumsey pamphlet wars in 1788. After studying their plans, he thought he could do better. On January 15, 1790, he submitted details for his inventions to Boston's Academy of Arts and Sciences, which found them commendable. He traveled to New York, where he petitioned Congress for patents and tried to drum up financial support. He explained his plans and showed models to President Washington and several other influential men, including John Stevens.[29]

Read's application asked for patents for improvements on the steam engine, a steam-powered land carriage, and a paddlewheel-driven steamboat.[30] He felt that Fitch and Rumsey had wrongly rejected paddlewheels (Read was either unaware or skeptical of Franklin's views on their supposed inefficiency). To make sure, he put the idea to a test. He built a small boat fitted with a hand crank that turned a pair of paddlewheels and took it out on Porter's River, near Danvers. He found the paddlewheels worked fairly well. In those early days of science and technology, direct experimentation was not put to use nearly as often as it should have been. Intellectual reasoning—or "thought experiments"—was considered a valid way to solve a problem, as evidenced by the many early patent applications submitted for devices that had not yet been built to full scale.

Read may have fallen into that category. It is not clear whether he ever built his proposed steam engine, with its new kind of multi-tubular boiler and an improved cylinder, but his ideas were revolutionary. Following the same reasoning that led Rumsey and Voigt to design their tubular boilers,

Read replaced the usual pot boiler with tubes, the idea being to expose a greater surface area to the fire. But instead of a single long, bent tube that exposed a surface of only 60 square feet to the fire, as Rumsey's did, Read proposed seventy-eight single tubes, each about 5 feet long, grouped vertically within a round container and exposing 195 square feet of surface to the fire.

Read knew that this smaller, lighter, and more powerful steam engine would make steam-powered land carriages feasible. His engine used high-pressure steam and eliminated the air pump, a condenser, and a working beam.[31] In a later version of his boiler, he reversed the usual arrangement and ran the flames and heated air inside the tubes, which heated the water outside them—a fire-tube boiler. He also redesigned the steam cylinder.[32]

Read's drawings for a four-wheeled steam carriage show a single boiler in the front center that would deliver steam directly to two cylinders, one for each front wheel. But when his plan for this early automobile was read aloud to the House of Representatives in February 1790, some of the members struggled to contain their laughter. Read was humiliated and wrote later that he was "stung by the indignity of the House."[33]

Sometime that year he happened across an old volume of the *Transactions* of England's Royal Society that described the use of paddlewheels to move a steam-powered boat in France, perhaps Jouffroy's, in the early 1780s. Read decided that under the patent act he could not legally patent a device that had already been invented, so he tossed out paddlewheels and devised what he called a "rowing machine," a chain-operated device that controlled a series of paddles. Although no drawings of it have survived, Read's device sounds oddly similar to Fitch's first design of 1785, which was never built and which he called an endless chain of paddles.

On January 1, 1791, Read sent in a new patent application to replace his first, deleting the steam carriage and substituting his rowing machine for paddlewheels in a steamboat.[34] Had Read not been so sensitive, he might have gone down in history as the inventor of the automobile. Had he not been such a stickler for the law, he would have caused problems for Fulton years later.

For reasons unknown—perhaps simply because investment money was scarce in those bad economic times—Read never found enough financial support to carry out his projects. In the late 1790s he built the Salem Iron Factory with several partners and received a patent for a nail-making machine. (Nail-making machines were a hot idea at the time; the records of the Patent Office show that in 1823, the office held models for ninety-five such devices, by far the most of any category reported.) He later served as a U.S. congressman and as a county judge. Before he died in 1849, he must have shook his head many times as he watched his ideas make fortunes for other men. Variations on his design for multi-tubular boilers could be found in the successful steamboats (using paddlewheels, of course) and locomotives of the era, beginning with Stevens and Fulton in the United States and George Stephenson in Great Britain.[35]

The fifth applicant, Isaac Briggs, soon withdrew his patent application for a steam wagon, for reasons unknown. It seems likely that this was the same Isaac Briggs who was a surveyor and mathematician from Brookeville, Maryland, and who later served as a surveyor general of the United States and as an engineer on the Erie Canal project. In 1788, a former New Jersey citizen named William Longstreet moved to Augusta, Georgia, and that same year received a Georgia state patent for a steamboat with Briggs, who had recently moved to nearby Wilkes County; this steamboat is said to have run briefly on the Savannah River. In 1802, Briggs and his brother, Samuel, Jr., took out a U.S. steam engine patent. This undoubtedly was the same Samuel Briggs who had made parts for Fitch's steamboat in 1786.[36]

Just two weeks before the February patent hearing date, Remsen sent the steam patent applicants another letter. He told them the hearing was postponed so that the commissioners could await the outcome of an amendment to the patent act.[37] Jefferson apparently was hoping to put off his confrontation with the steam inventors on the chance that Congress would move quickly and get him off the hook. Less than a year after the patent act became law, the applicants and the commissioners had grown completely frustrated with it. Because of the time it took to carefully examine each invention, fewer than half of the patent applications were being approved—only fifty-seven applicants would receive patents in the almost three-year run of the first patent act. Jefferson himself had drafted a bill to fix these problems.

Fitch in particular was fed up with the seemingly endless delays. In a snit, he showed up at the State Department and demanded to see the commissioners. He later wrote that he managed to get an audience of some kind, although it is not clear who, as his first biographer, Thompson Westcott, describes: "He appealed to them in virtue of his distresses, and urged that he was kept in idleness and suspense until their decision should be made. 'I showed them,' said he, 'all the clothes I had in the world, except a few old shirts, and two or three pair of old yarn stockings, all in darns, like those which I had on, that they could see I was there all in rags.' Appeals like these were useless, the Commissioners were not to be affected by the presence of a poor, wretched genius, whom they knew was derided as a madman."[38]

To Jefferson's certain dismay, Congress adjourned on March 2 without acting on the patent amendment. Fitch, though, was relieved because he felt certain that under the current law the first steamboat patent was his. The law specified that a patent be given to the "first and true inventor," and he could prove without a doubt that he had publicly laid out his plans well before anyone else. Impatient to get on with it before the rules changed, Fitch continued to show up at Jefferson's door until, as Westcott writes, "to get rid of the persevering pest, the Commissioners appointed the first Monday of April to hear the steam–boat cases."[39]

On April 4, Fitch arrived at the State Department offices for the hearing. Joseph Barnes, representing Rumsey, was also present, but there is no indication that any of the other steam applicants attended. Fitch led off the meeting by eliminating Stevens from contention. He told the commissioners Stevens had offered to support him in 1786, well before Stevens began to think about steamboats himself, proving without a doubt that Fitch was first with the idea. Fitch then argued that Read clearly wasn't a contender because he first presented his ideas publicly less than a year earlier. Of course, Fitch was assuming that the board would make its decision based on which man was the "first and true" inventor, even though no rules existed to guide the commissioners if two or more applicants claimed to be the first inventor.

It wasn't long before things got interesting. Barnes and Fitch rehashed their usual squabble, with Barnes claiming that Rumsey *thought* of steam in 1784. Fitch countered that the timing of a thought couldn't be proved but that actions could—his petition for steamboat rights to the Continental Congress in 1785 was part of the public record. The two men bickered back and forth for a considerable time. The commissioners sat back and listened, probably looking at their watches every so often. Finally, Barnes proposed that he and Fitch submit to arbitration, with each side choosing three men of character to judge the case. Fitch heartily agreed and named six possibilities, including four well-respected men who happened to be long-time supporters—David Rittenhouse, Andrew Ellicott, and University of Pennsylvania professors Dr. John Ewing and Dr. Robert Patterson. Barnes suggested several others, three or four of whom were members of the Rumseian Society. Each time a name came up, the other man would knock it down. After much argument—the meeting dragged on into the following day—Barnes withdrew his suggestion for arbitration. The commissioners sighed and set a time for another hearing to take place two weeks later.

Three days before the April 22 hearing, Fitch completed his masterpiece of defense, a fifty-six-page memorial he had been composing for months, and dropped by the State Department office to deliver it to the commissioners.[40] The clerk, Remsen, was away that day, so Fitch insisted on seeing Jefferson. The secretary of state made it clear that he did not want to touch the document until it had been properly recorded. But Fitch would not budge and insisted that Jefferson at least read it. Jefferson finally consented, probably knowing it was the only way to get rid of him. It must have taken a while to slog through it, if he actually did; the first twenty-nine pages consisted of facts and arguments, followed by sixty-two exhibits and details about the weak claim of Stevens and others. A few hours later Jefferson returned the document to Fitch and told him to file it with Remsen when he returned.

The hearing day fell on Good Friday, and to Fitch's frustration the board cut the meeting short so that one of the three commissioners, Attorney General Edmund Randolph, could attend church. They reconvened the next day, when Jefferson sidestepped the issue of who deserved the first patent. He ruled that all steamboat patentees should be awarded patents dated the same day. Here is Fitch's account of the proceedings, written shortly afterward in a letter to Thornton:

> I presented my petition to the Commissioners. Which was read and laid on the table without one single observation on it. . . . Knowing it to be the order of the Board that patents should issue to all [rival] applicants [I] then requested that the oldest patent might be granted to me [so] that I might become the defendant in any suit hereafter brought as I could prove to them that Rumsey was twelve months posterior to me. . . . Mr. Randolph said that the oldest applicant must have the oldest patent, thinking as I suppose that Rumsey was the oldest as he was before me in his application to them. To which they all seemed to assent. I discovering what appeared to me partiality spoke in a dry determined and positive manner uncommon to myself and informed them that I was the first applicant and that I had communicated the plan to Congress in August 1785 and [had] in March 1786 petitioned to them for an exclusive right for the same. . . . To which they all appeared silent for some time, when Mr. Jefferson said that they could make no distinction in the patents nor give one the preference of another. To which I warmly remonstrated against but to no purpose when he said and persisted in it that it would make no odds to me. Which I could not then nor can I yet see the propriety of that assertion. And told him I could not conceive it in that light but as I was unacquainted with law I might probably be mistaken. Finally they ordered patents be issued to all and for every mode and to bear even dates with each other. I did not like to tell him that he just said that there was a preference in the oldest patent. . . .
>
> At any rate the change in their judgments was very sudden and no other reason given for it than my making it appear I was the first applicant. Which seemed [to them] to be a very good reason why the first applicant should not have the first patent.[41]

The commissioners' decision was final and could not be appealed. Patents were granted to all applicants—Fitch, Rumsey, Read, and Stevens—that day. They weren't officially issued until August 26 because Washington's signature was necessary and he was out of town. In the meantime, the commissioners instructed Remsen to rewrite each man's specifications in such a way as to make them distinctive. He did so, but was not able to get around the problem caused by giving rights to both Fitch and Rumsey for propelling boats by air- or water-jet propulsion.[42]

Fitch was furious. He stewed for three days, finally deciding that the only way to keep the project going was to head west, to establish a company in the Mississippi valley, or at the very least to sell his Kentucky land to generate much-needed cash. He sent a note to the few remaining members of the Steamboat Company telling them he would like to speak to them that evening.

> To the Steam Boat Company—Gentlemen: . . . The uncandid feelings of our nation toward me has determined me for the present to quit a scheme of the first importance both to you and to our engine with a view of reassuming it again when the justice of our country will protect our laudable endeavours. . . .
>
> We have worried against wind and tide long enough and there is little doubt with me but the tide will turn in our favour in the course of one year. In which time permit me to go to Kentucky for and in behalf of the Company. If not, I shall go on my own account and see whether I can do anything toward perfecting the scheme or not.[43]

The congressional efforts to change the 1790 patent law continued for nearly two more years. Jefferson wanted to eliminate examination and replace it with a registration system like that used in England. Jefferson spent hours each month evaluating patents. He wrote to his friend Hugh Williamson a year later, "Above all things he prays to be relieved from it [Jefferson's duties as patent commissioner], as being, of everything that was ever imposed upon him, that which cuts up his time into the most useless fragments and gives him from time to time the most poignant mortification."[44]

In his draft bill, Jefferson proposed that a patentee would receive a patent certificate so long as he completed the paperwork, submitted a model, and paid the fees. Jefferson also proposed that before the patent could be granted, the inventor would have to file all three documents in the clerk's office of *every district court in every state* of the United States. That wasn't all. The inventor would also have to publish these documents "three times in some one gazette of each of the said districts." Jefferson's reasoning behind these onerous requirements was to inform the public about the invention and to "warn others against interference therewith."

It is hard to imagine that Jefferson wasn't trying to discourage all but the most serious (and well-off) inventors. On February 7, 1792, the House committee presented its bill, probably Jefferson's draft, but it was tabled. Naturally, the steamboat applicants were aghast when they read its contents. Nathan Read immediately complained to Congress, railing against the "extraordinary precaution of advertising the discovery before the patent is granted" and explaining that others could steal the idea from the original inventor before a patent was issued, if the specifications were detailed enough. The Senate later removed the advertising requirement, perhaps thanks to Read. Fitch, not surprisingly, was also unhappy. Three days after the bill was presented, he delivered a petition to the House saying the law would injure him directly. He told the committee he "had no idea that he must go all the way from Kentucky to Cape Cod, and quite the distance of the province of Maine, to publish his inventions, and to pay out large fees wherever he goes for the same."[45]

Barnes, Rumsey's attorney, protested another aspect of the proposed bill. In a pamphlet he published in Philadelphia in 1792 he wrote that the bill "contemplates, at the expense of the American genius to import European arts and literature!!!" He was referring to a clause that would use the fees received from the inventors for "procuring and importing such useful arts and machines from foreign countries." He was glad to see action underway to replace the 1790 act, though, calling it "worse than none."[46] He argued that any future law must deal with the issue of priority, because the present law, which gave two inventors patents for the same device, places an unfair burden on the inventors by forcing them to resolve their disputes in court.

Nearly a year later, the Patent Act of 1793 became law on February 21. Although the fourteen-year term and the requirements for novelty, specifications, and models remained the same as under the old act, almost everything else had changed. Jefferson won on at least one count: The new law replaced the time-consuming examination process with a system of registration. (Patent examination wouldn't return until the Patent Act of 1836, after years of problems of a different kind caused by the registration system.) The applicant would still pay a fee, but now it was a steep $30, about six times the previous amount and equal to about one-third of the average annual per capita income in those days. Also, anyone holding state patents granted before the Constitution was ratified would have to surrender those patents in order to obtain a U.S. patent. Basically, anyone who met the administrative requirements and swore an oath that his or her idea was original would receive a patent.

Because the Rumsey-Fitch priority battle was still fresh in the lawmakers' minds, a clause specified that if two patent applicants claimed the same invention, the person who was first to invent, not first to file, would receive the patent. Since this wasn't always easy to determine, a three-person arbitration board would be formed, consisting of one member named by each patentee and the third member chosen by the secretary of state. No appeal of that board's decision would be possible. Disputes arising after a patent was issued would be decided by the courts.[47]

Jefferson was vastly relieved to be free of his patent duties. As a letter to a friend many years later revealed, he never really believed that patents were of any value:

> He who receives an idea from me, receives instruction himself without lessening mine; as he who lights his taper at mine, receives light without darkening me. That ideas should freely spread from one to another over the globe, for the moral and mutual instruction of man, and improvement of his condition, seems to have been peculiarly and benevolently designed by nature. . . . Inventions then cannot, in nature, be a subject of property. Society may give an exclusive right to the profits arising from them, as an encouragement to men to pursue ideas which may produce utility, but this may or may not be done, according to the will and convenience of the society.[48]

The first patent issued under the Patent Act of 1793 was to Eli Whitney for his cotton gin. Judging by the time lag between application and award, the new process didn't appear to be much of an improvement: Whitney applied on June 20, 1793, and received his patent certificate the following March. Because his device was fairly simple to copy, he spent a great deal of time—and most of his profits—suing infringers in court. But he was one of the lucky few of the early patentees to make money—more than $90,000 in royalties over the term of his patent, according to one estimate. Had they known of Whitney's success, Fitch and Rumsey surely would have been astounded.

EIGHT

"ALL FURTHER PROGRESS IS IN VAIN"

Pioneering don't pay.

—*Andrew Carnegie*

S hortly after Jefferson gave Fitch and Rumsey patents dated the same day, most of the remaining Steamboat Company shareholders quit. Still supportive and involved from afar was William Thornton, the majority shareholder and "vivifying spark that gives life and animation to the steam boat," as his bride, Anna Maria Brodeau, put it.[1] The newlyweds had left Philadelphia the previous fall for an extended visit to his family's plantation on Tortola. The handful of shareholders who had hung on—Richard Stockton, Edward Brooks, Jr., and Dr. Benjamin Say—spent a good part of May 1791 quarreling over what to do next. Stockton took charge of the company's activities in Thornton's absence. He was twenty-seven years old, a Princeton-educated lawyer and later U.S. senator whose father, also named Richard, was a signer of the Declaration of Independence.[2]

Fitch was not enthusiastic about carrying on. Unless word arrived that Rumsey was successful in England and was coming back to America, he saw no signs of competition on the horizon. Moreover, he was angry. He believed, incorrectly, that his patent would prevent others from building steamboats during its fourteen-year term. Bitter over Jefferson's decision, he planned to go on strike, to punish his ungrateful and shortsighted nation by depriving it of the steamboat.

Three days after the patent decision, Fitch had called an emergency meeting of the Steamboat Company, sending a note to the directors begging them "to relieve their unfortunate . . . John Fitch."[3] But the directors voted to continue. To raise funds for a new boat, they decided to sell off parts from the old boat and chip in more of their own money (Thornton sent £100). They convinced Fitch to stay in Philadelphia through the summer, promising to pay his back wages. Their plan was for Stockton to take the *Perseverance II* (presumably fitted with sails) down the Atlantic coast, into the Gulf of Mexico, and up the Mississippi by late summer. That July, Stockton would receive permission from the Spanish governor to take the steamboat up the river at New Orleans, so long as it carried no cargo.

Fitch resumed work on the boat in late May but only after much disagreement among the directors on how to proceed. Technical problems plagued Fitch and his mechanics all summer. When they tried the new engine for the first time in September, "the water gushed out in streams as thick as my fingers in many places [so] that I could not raise force of steam sufficient to move the works in the first place. And it being a large engine we could not move the great piston by hand as we could in the small to put the works in motion. . . . It may be easily conceived that my anxieties of mind have not been trifling." Cost-cutting measures on the part of the directors—for example, using wood rather than copper for the boiler and using lead instead of a stronger metal for the engine—were most probably to blame for the failures.[4]

Thornton had warned Fitch about using a wooden boiler earlier that spring. "I beg you to take out a patent for the boiler. It is the very first ever made with copper in that way, and they can do nothing without it. Wood will not answer."[5] He was full of more advice in a June letter: "I think your condensing water cannot be supplied in too abundant a quantity, your last water pump was very defective. You will remedy the next, and give as much water as can possibly be wanted. What do you think of four paddles at the end instead of three large ones? For while one is going down and the other rising the third has scarcely obtained its action . . . If the axis were lower, and the paddles more numerous it would have more the effect of a wheel. . . . Perhaps Padre Stockton has examined all these things before—if not, give them a moment's thought."[6]

Thornton's mother-in-law, Ann Brodeau, a bright, attractive widow who ran a girls' boarding school in Philadelphia, kept him informed of the latest news, including progress reports on the steamboat. She often visited Fitch on the waterfront. Writing to Thornton in late September, she describes how she went down to the wharf to see the steamboat, only to find that the workmen had left for the day. "It did not appear to be in a state of great forwardness. I fear it will not arrive at the Mississippi this fall if ever. I call'd on Mrs. House who expresses much kindness towards you. She said she heard them say that the old boiler would not do, but that they expected it would be finished and that Mr. Stockton intended to go to New Orleans in it."[7]

Thornton had moved to his family home in Tortola to practice medicine, in hopes of making some real money now that he was married. But there was little demand for his services, so he spent his days pursuing a variety of interests. Something of a Renaissance man, he collected botanical drawings for a natural history of Puerto Rico he planned to publish, he wrote a treatise on the English language called *Cadmus* that later won an award from the American Philosophical Society, and he sketched designs for the planned U.S. capitol building in Washington. He also continued to work on a plan, begun in Philadelphia, to establish an independent settlement in West Africa for freed slaves.[8]

Ann Brodeau's prediction about the steamboat proved true. Problems kept the boat tied up that fall, but Fitch continued to work on it as long as weather permitted. One day in early October, as he readied the steamboat for what he hoped would be its last trial of the year, he heard a familiar voice calling to him from the dock. He looked up to see his old friend and partner, Henry Voigt. He knew what was coming. Voigt was back to taunt him about his continuing string of failures. Fitch turned back to his work. As if to prove Voigt right, that day's trial was another failure, this time blamed on a faulty air pump.

A few days later, the company reluctantly agreed to spend a few more pounds on a new air pump. The news reached Voigt and sent him straight back to Fitch. He showed up at the wharf "early one morning, seemingly in a passion and demanded that [I] should immediately go off to the Ohio to put the other boat in practice or he would agree with another to do it, as there was a gentleman very solicitous of it. I replied thus: 'Harry, if you can make a good bargain with him, do it. There is business enough for a dozen men on the Ohio.'" Voigt replied with a final insult, yelling to Fitch as he walked away that the new boat would never work without his help.[9] Voigt's antagonism can be traced to anger and jealousy, professional and personal. He never felt Fitch had given him enough credit for the steamboat's successes, limited though they were.

Ann Brodeau visited Voigt around that time to see his model for a horseboat, his latest project. She described it to Thornton: "To me there appears many inconveniences. If it is meant simply for passengers, four horses, and forage for them, will take a great deal of room and be disagreeable companions, and it cannot be meant to carry merchandise as the boat is to consist of two linked together and the depth in the water no greater than a canoe."

She continued, "From the foregoing I suppose you will guess the fate of the steam boat. I told him [Voigt] who I was, and asked him to give me an account of the steam boat as I was going to write to you and wished you to be informed how they were going on, and I did not know where to find Mr. Fitch. He gave me a long account of their proceedings, but told me if I would enclose it to you he would write an amplified account of it, for it was a shame that gentlemen who had contributed the money and were absent should be imposed on."

Mrs. Brodeau waited a few days, even dropping by Voigt's house once to find him away, but she never received his report. She told Thornton that he may have decided against putting his thoughts in writing "lest he might hereafter embroil himself with Mr. Stockton as he lays the greatest part of the blame of the steam boat's failure on him. He says that he differed in opinion with respect to the construction of the new boat and declined having anything to do with it, as he foresaw its failure from the beginning. . . . He says moreover that Johnny Fitch has not been so much to blame as Mr. Stockton."[10]

Thornton wrote back, "I lament exceedingly the fate of the steam-boat, but mean to expend no more money upon it. If I had been present the foolish schemes they pursued of wooden boilers and such trash would have never taken place. The horseboat as you well observe is only another whim, for whenever water is rough although the horses are tied in vessels they are dangerous by running back always to leeward, and a boat could never be worked by horses in dead calm. 'Tis a scheme that will never answer!"[11]

Animal-powered boats were nothing new; the idea had been around for centuries. In the most common arrangement, farm animals tied to horizontal poles walk treadmill-like on a circular platform that drives a post, or whim, which turns a wheel or paddles. Fitch had come up with a horseboat design, featuring his crank-and-paddles device, in 1787. As he later explained to a skeptical Thornton, Fitch had wanted to put the federal patent for the steamboat in both his and Voigt's name, but the company directors disapproved, saying that Voigt was "an obstinate Dutchman and would give them much trouble."[12] Fitch did not feel right about this because he knew Voigt's contributions were crucial. In an attempt to give Voigt a consolation prize, Fitch agreed to let him patent the horseboat in his own name, even though he felt the invention was rightfully his.

Voigt received a U.S. patent in early August and immediately began raising funds to build a boat.[13] He was soon shocked to learn that Fitch was considering partnerships to build horseboats in America and Germany. Voigt confronted Fitch and angrily questioned him about his actions. Fitch pointed to an agreement the two had signed stating that they would share equally in their endeavors. As Fitch saw it, the horseboat patent was half his. Voigt disagreed, telling their friends and supporters that Fitch was infringing on his patent. The horseboat incident led Thornton to distrust Fitch for the first time.

Voigt also felt that Fitch had ruined his relationship with Mary Krafft. After the birth of her second child by Voigt, she began calling herself Mrs. Fitch. She and Fitch were now living together at her tavern, with the baby, as man and wife, an arrangement that had apparently moved into the bedroom. Fitch writes, "Thinking [that] by acting as her most intimate friend I possibly might wean the affection of her from the man who had treated her so basely, of course I at that time became more intimately acquainted with her than ever I had been before with a design of making her my lawful wife as soon as I could be persuaded her affections were so far called off from him. . . . He [Voigt] being sensible of my determination became much enraged as if I had debased him from his choicest pleasures."[14]

For reasons probably having to do with the continuing problems with Voigt, Fitch encouraged Mary to leave town for an indefinite period. That fall, she made plans to move to New York, but she told Fitch she wanted to see her young son before leaving. Fitch promised to locate the child. Voigt evidently knew who was caring for the baby but "although I had demanded it in very positive terms from him," Fitch could not get Voigt to talk. More than a year earlier, Fitch had carried the hours-old infant to a nurse. He then visited the child weekly to keep Mary informed. When the child was about a year old, someone took him from the nurse and Mary became concerned.

On November 1, Fitch wrote Voigt a letter on the matter, concluding with a threat: "This, sir, I demand as a right of a friend to know and if I am again refused be assured that I shall call for him in a serious way before a justice of the peace for I think her requests are but reasonable and am resolved to grant every reasonable request of hers. I hope your obstinacy will not carry you to the disgrace and perhaps ruin of both your families but call and see us at the shortest notice and take a walk with us that she may see the object of her desire."[15]

A messenger returned the letter to Fitch unopened, saying that Voigt wished to visit them that evening. Fitch told the messenger to tell Voigt that they would not see him until he had read the letter. A short while later, Voigt came to the tavern, and Fitch, questioning him at the door until he was convinced he had read the letter, let him in. That was a mistake. "As he was then drunk and kept so all the next day the abuses which he gave to her and me ought to remain unpardonable, but I forgave him the abuses to myself for which he took me around the neck and kissed me. The offense he gave to my other friend has not been settled, neither can I see any reparation can be made for the loss of character which he in his wild extravagance and very falsely as well as loudly asserted to our listening neighbors."[16] Voigt had shouted for all to hear that the baby was his, not Fitch's. Fitch and Mary were mortified. Not long after, Fitch learned that little John Barney Voigt had died of a sudden illness in his new home.

Voigt's harassment continued. A few days later, Fitch made Voigt sign a witnessed statement: "I the subscriber promise to keep peace with John Fitch, not to molest his person nor property in no way whatever, I say to keep this promise for the space of three months, and for the performance thereof, I find myself in the sum of one hundred pounds lawful money of Pennsylvania, proviso nevertheless, the said John Fitch shall not give any case of anger to the said Henry Voigt in any manner whatever."[17]

In December, as planned, Mary left for New York, leaving her children with friends. The bitter fight between Fitch and Voigt over horseboat rights continued into early January, when Fitch forced Voigt to sign another witnessed statement, this one acknowledging that Fitch was the original inventor of the method of rowing by cranks and paddles.[18]

A severe winter kept Fitch from working on his boat and engine, so the company stopped his pay until work could resume in the spring. The only good news an impoverished Fitch had that winter was a letter from Aaron Vail saying that his French patent for the steamboat had been issued. Vail asked Fitch to come to Lorient as soon as possible to begin building the boat.[19] The letters of Fitch and other Steamboat Company members during that period indicate that he planned to leave for Europe that spring, but it seems the chance to finish the *Perseverance II* kept Fitch in Philadelphia. He supported himself with silver-smithing and clock-making work, but just barely. With time on his hands, he continued to work on his nearly three-hundred-page "Steamboat History," which he had begun in 1790, and started writing the story of his life.[20]

Sometime late that winter, Mary wrote Fitch from New York to inform him she was coming back. This news prompted an immediate reply:

Dear Nancy [his nickname for Mary], Consider now that you are on ticklish grounds and that you rather ought to act from reason rather than from inclination. I conjure you to take this one advice of mine, to send orders to Esquire Weaver to send a certain part of your effects and things to Charleston and get into some small way of business and live as becomes a good citizen, which I have no doubt you would, and save yourself and your family, or if you wish to go to Europe I now offer to take you under my care but should expect your strict compliance with my advice in every case. Dear Nancy, do not be offended at my plainness, you will know I have frequently used it before and whatever is good make use of it and what is not esteem it as coming from an honest heart and a sincere friend.[21]

He told her that her children were all doing well, including her second child by Voigt: "Harriot was here a few days ago, well and hearty as many of our neighbors can testify."[22] Fitch evidently wrote another letter to Mary a short time later, for she wrote him back immediately: "I received yours yesterday with the dreadful news of the loss of my dear babe." Little Harriot, the child that Fitch was ready to call his own, had died. Mary wrote a note to her lawyer, Weaver, instructing him to "please to let my husband Mr. John Fitch as much money as he will ask you for to pay the funeral charges of little Harriot Fitch." In her letter to Fitch, she pleaded that he not to go to France—"they will get out of you what they can then you may go to hell, there is bread plenty for you in your own country." She signed it, "your affectionate wife, Mary Fitch."[23]

Against Fitch's wishes, Mary returned to Philadelphia. He could not bear to continue such a painful relationship. He moved out of her tavern, leaving behind a curt note saying that she had lost the most sincere friend she had ever had, and adding a postscript: "I do not wish to hurt your feelings but wish you to have some serious reflection."[24] Fitch took a room in the boardinghouse of another widow, Hannah Levering, on Front Street. In April he paid Mary his back rent in full.

Fitch wrote to Thornton, who was still in Tortola, in late March 1792 with the news that he was back at work on the boat and hoping to complete it soon.[25] Fitch's account ledger for April 1 shows plenty of activity—payments for boards, nails, twine, and other supplies, as well as three intriguing entries for several pints of spirits and beer for the services of a Mr. Evans.[26] This was probably Oliver Evans, who was becoming well known for his automated grist mills. Evans had recently moved to Philadelphia, and his long interest in steam-powered vehicles surely would have sent him straight to Fitch's workshop. Many years later, in a court case, Evans said he had once advised Fitch to use side-mounted paddlewheels, a suggestion Fitch rejected. Fitch may have failed to heed Evans's advice on steam engines as well, because after more than a month of tinkering, the engine was still not working right.

On May 6, Fitch wrote a letter to Dr. Say and Mr. Brooks in which he concluded that the biggest problem with the steam engine was the size and loading of the piston. Blaming Voigt and a mechanic named Matlack for the problem, Fitch calculated that a larger wheel would solve the problem. He asked for $4 to pay a carpenter to do the job quickly, adding that he hoped they would "indulge me in this trifling request." They turned him down.[27]

Fitch was back to his usual penniless state. He was forced to do what he hated most: beg for loans to finish the boat. On May 23, he signed a note promising to pay back his former Kentucky landjobbing partner, Daniel Longstreth of Warminster, £100 in current Pennsylvania money.[28] On June 29, he wrote to his old supporter David Rittenhouse asking to borrow £50, which he planned to use to go to Kentucky, sell the land he owned there, and repay him with £100. Rittenhouse turned him down. On July 6, he offered the same deal to Richard Wells, a former Steamboat Company director. Wells declined but lent him £10.[29]

"Often I have seen him stalking about like a troubled spectre, with downcast eye and lowering countenance, his coarse, soiled linen, peeping through the elbows of a tattered garment," recalled a Philadelphian named Thomas P. Cope, who wrote of Fitch many years later.[30] Cope described how, around this time, Fitch would frequently visit the Kensington workshops on the Delaware waterfront, where he and Voigt had built their steamboats. There he would talk with two of his former mechanics, Peter Brown and John Wilson. Cope related this story:

> Fitch called to see him as usual—Brown happened to be present. Fitch mounted his hobby, and became unusually eloquent in praise of steam, and of the benefits which mankind were destined to derive from its use in propelling boats. They listened, of course, without faith, but not without interest, to this animated appeal; but it failed to rouse them to give any future support to schemes by which they had already suffered. After indulging him for some time in this never failing topic of deep excitement, he concluded

with these memorable words: "Well, gentlemen, although I shall not live to see the time, you will, when steamboats will be preferred to all other means of conveyance, and especially for passengers; and they will be particularly useful in the navigation of the River Mississippi." He then retired; on which Brown, turning to Wilson, exclaimed, in a tone of deep sympathy, "Poor fellow! What a pity he is crazy!"[31]

By midsummer it was apparent that Fitch was considering ending his life. In mid-July he asked three friends to witness his will, and over the next few months he wrote letters that had the intent of wrapping up his affairs. He wrote to his son-in-law, James Kilbourne, in Connecticut, asking him to pass along word of his love for Lucy (the daughter he had never seen) and her older brother, Shaler, who was a toddler when Fitch walked out twenty-three years earlier.[32] He wrote a letter to his landlady, Hannah Levering, telling her that he would be giving cash to a certain attorney "before my death" to "give you as little trouble as I can in collecting what is your just due." He told her who owed him money and how she could collect it and instructed her to pack up his loose papers and maps and deliver them to his executors.[33] He wrote to the son of Israel Israel, a friend and former share-holder, saying that he was leaving him the Pennsylvania flag that he had received for his first steamboat and suggesting that the young man should court and perhaps marry Mrs. Levering's daughter (followed by a postscript asking that if he did marry her, that they name their first son after him.)[34]

Fitch also completed his history of the steamboat and his autobiography that summer. His plan was to give these writings to the Library Company of Philadelphia for safekeeping so that future generations would know his story. Fitch was still angry at Jefferson for refusing to give him the first steamboat patent. He believed that with a priority patent, he would have had plenty of investors and would by now be building steamboats for the Ohio and Mississippi.

As he packed up his papers for the library, he decided to give Jefferson a chance to review what he had written. He drafted a letter to Jefferson that began with these words: "I, sir, am sorry to live in a state that no sooner becomes a nation than it becomes depraved. The injuries which I have received from my nation, or rather from the first officers of government, has induced me, for a lesson of caution to future generations, to record the treatment which I have received, which will in a very few days be sealed up and placed in the Library of Philadelphia, to remain under seal till after my death, in which, sir, your candor is seriously called into question."[35] But Fitch never mailed Jefferson the letter, talked out of doing so by his friends.

A few months later, on October 4, he took his papers to the Library Company, accompanied by a letter instructing the librarian to keep them sealed for thirty years. He also asked that any future publisher put them in proper order and make the necessary grammatical corrections. Three weeks later, he wrote the librarian a second letter that cited two exceptions to his thirty-year seal. One allowed the documents to be opened if someone came forth who was willing to edit them and publish a thousand copies. The second was a dagger aimed at Jefferson: "This is further to request you, that should Mr. Jefferson ever be aiming towards the president's chair, by all means to obtain leave to break the seals and extract what affects the commissioners of Congress, and then seal them again."[36]

As Fitch sunk deeper into poverty and depression, Voigt's life began to improve considerably. His old employer, David Rittenhouse, was sworn in as the first director of the U.S. Mint on July 1, and a few days later President Washington appointed Voigt as chief coiner, a prestigious job he would hold until his death in 1814. The mint had been established by an act of Congress that April, and Rittenhouse was fortunate to have a man with mint experience fill this highly technical position. In his application in April, Voigt noted that he knew "how to use every engine belonging to a mint, but [could] make every one himself in all its parts complete (except engraving the dies)—and even [had] made some improvements in the machinery whereby a considerable expense was saved."[37] By that fall, with imported machinery, Rittenhouse and Voigt had the mint up and running in a new building at Seventh and Arch Streets.[38]

Around this time, Fitch wrote a poem summing up his anguish:

For full the scope of seven years
Steam boats excited hopes and fears
In me, but now I see it plain
All further progress is in vain

And am resolved to quit a scheming
And be no longer of patents dreaming
As for my partners *damn them all*
They took me up to let me fall
For when my scheme was near perfection
It proved abortive by their defection
They let it stop for want of rhineo
Then swore the cause of failure mineo.[39]

The written record fades at this point, but it appears that Fitch was pulled from the brink by the horseboat. In September, Fitch and a man named Elmer Cushing of Montreal, Canada, signed a contract in which Cushing would receive half of Fitch's patent rights to build a horseboat to operate on the Delaware between Philadelphia and Trenton or Bordentown.[40] This endeavor apparently never got off the ground. The agreement specified that Cushing begin building a boat by the first of December. When that did not happen, John Nicholson, a leading Philadelphia financier and Pennsylvania's comptroller general, became interested in the idea. On December 28, Fitch sent a note to the Steamboat Company members asking them "by desire of Mr. Nicholson" to meet at Mr. Israel's house that evening "on business of importance." Things moved quickly after that. Papers were signed on January 4, 1793, giving Nicholson half of Fitch's patent rights and employing Fitch as superintendent; Nicholson agreed to pay all costs.[41]

Work began immediately. Fitch's account book for January shows a steady list of parts being ordered and paid for, then a note complaining of a two-month delay in the construction of the boat. Now that Philadelphia was the nation's capital, a building boom was going on, and carpenters and mechanics were in high demand. Fitch's last account entry was for February 1: "The iron works delayed."[42] Work on the horseboat came to a halt, partly because of supply problems but probably also because Nicholson was in trouble: He was being impeached by the Pennsylvania senate, charged with trafficking in illegal stock certificates and misappropriating state funds.[43]

Revitalized by the horseboat experience, Fitch decided to grab the only opportunity left: to build a steamboat in France with Aaron Vail. The circumstances surrounding Fitch's sudden departure are not recorded, but he left Philadelphia on February 14, 1793, for New York, where he boarded a packet ship for France. Two days before he left town, he gave Thornton, who had by then returned to Philadelphia, power of attorney to settle his accounts and act on his behalf.[44] He also recorded his observations on navigation in general and horseboats in particular for Nicholson later that month, in case work resumed in the future.[45]

As Fitch sailed away from the docks that had brought him so little triumph and so much grief, he must have breathed a deep sigh as he looked one last time at *Perseverance II,* pulling at her moorings. He could not have known that across the Atlantic, Rumsey's *Columbian Maid* had just made her last run.

NINE

LEECHES AND SHARKS

Soon shall thy arm, UNCONQUER'D STEAM! afar Drag the slow barge, or drive the rapid car; Or on wide-waving wings expanded bear, The flying-chariot through the fields of air.

—*Erasmus Darwin, 1791*

Twice Rumsey was told by a messenger that his lordship, the earl of Stanhope, had shown up at the wharf wanting to see his steamboat, sometime in the winter of 1789–90. And twice Rumsey ignored this bit of news, certain that this so-called earl was really a London bailiff looking to haul him off to debtors prison.

A few days later Rumsey, hiding out in his room, got word for the third time that Stanhope had come by the wharf, demanding to see him. Judging by the man's dress and haughty manner, Rumsey's informant said he was fairly sure the man was not a bailiff. Rumsey, still skittish, asked George West, his close friend and fellow American who was a student at the London art school of Benjamin West (no relation), to give the man a tour of the *Columbian Maid.*

West took Stanhope aboard the boat and explained its workings as best he could. But Rumsey was right to be skeptical, albeit for a different reason. He learned a short time later that Stanhope had applied for a steamboat patent quite similar to the English patent Rumsey received not long after his arrival in 1788. "I have stopped him," Rumsey wrote, "and we are to have a hearing I expect soon before the Lord Chancellor. How it will end I cannot tell," he wrote to his brother-in-law back in Shepherdstown, Charles Morrow.[1]

Charles Mahon, the third earl of Stanhope, was an inventor himself and was quite interested in steam power; he would later correspond with Fulton and design a steam-powered printing press. As a politician and brother-in-law of William Pitt the Younger, then prime minister, he was well known for his liberal beliefs and wild ideas. One story goes that he invented a steam-powered land carriage that had the amazing tendency to speed up hills, stall on level ground, and stop completely on downhill slopes.[2] At the moment he was determined to launch steamboat service between England and France, in the hope that increased economic ties would prevent wars between the two nations. He had approached Boulton & Watt, but they refused to talk to any more steamboat projectors; Watt continued to believe that steamboats would never work well enough to be of practical use.

Rumsey, still smarting over the failure of his first steamboat trial on the Thames earlier that winter, was staying afloat financially through loans from friends. When one of those friends declared bankruptcy, the man's creditors combed the streets looking for Rumsey to pay up—"they were as inexorable as devils."[3] He felt he had to delay further experiments, fearing that if he failed again, all of his own creditors would descend on him in a heartbeat. As it turned out, he was idled anyway because his various suppliers and parts makers refused to take his orders on credit.

As if things weren't bad enough, Rumsey was told that if he accepted money from the Rumseian Society, he would void his English patent. When this news reached his supporters in Philadelphia, the treasurer of the society, Benjamin Wynkoop, "drew bills upon me to the amount of one thousand pounds, sterling." Rumsey couldn't pay up, so his name was put on a list at a public office: " . . . this was a new draw-back upon my reputation."[4] More likely, Wynkoop had learned that after more than two years and a hefty investment, Rumsey's latest attempt to launch a steamboat on the Thames had failed miserably. Society members back in Philadelphia were probably still shaking their heads in disbelief over Rumsey's rejection of a deal with Boulton & Watt.

One small hope kept the inventor going, and that was the chance that two acquaintances "of very philosophical minds" would follow through on their interest in forming a partnership. This deal was made in short order, and Rumsey signed a contract with American businessmen Samuel Rogers and Daniel Parker on March 25, 1790. Rumsey gave up a two-thirds share of his English patent in exchange for £1,000 in cash and £1,000 in goods. The goods would be shipped to Rumsey's brother-in-law Charles Morrow, who would sell them at his shop in Shepherdstown. Once the boat was built, Rumsey would receive a final payment of £5,500.

Rumsey had known Parker for nearly a year; it appears they met while both were traveling to Paris the previous spring.[5] Parker was a supplier to the Continental Army during the war, and in 1783 he supplied George Washington's personal account.[6] Shortly after the Revolution, Parker and Robert Morris, the Philadelphia financier, were the primary partners in a venture involving the first American trade ship to sail to China, the *Empress of China*. The cargo consisted of two items of special value to the Chinese: tons of North American ginseng and $20,000 worth of Spanish silver dollars. Parker lightened the ship's load of silver specie by more than $2,300 just before it sailed, telling the captain that he would repay the sum later and to keep his mouth shut. Parker never did, and when the other partners later learned of the fraud, he took off for Europe.[7] Little is known about Rogers, but Rumsey later wrote that he "proved to be a man of more cunning and duplicity that any man I ever met with." But he found Parker to be always "honourable."[8]

The shipment of dry goods left London for the port of Baltimore in April, but Rumsey received just £800 from Parker and Rogers, who claimed it was all they had on hand at the time. With great relief he wrote to Morrow about the deal, saying, "If I had not made the sale, I must have been in a London jail before this time."[9] But his relief was short-lived. In a letter to his one-time Virginia partner James McMechen, Rumsey wrote that " . . . unfortunately the whole of that sum fell into the hands of that scoundrel, Wynkoop."[10]

The idea behind sending English products to Morrow was to raise funds to repay Rumsey's debts in America, primarily to the Rumseian Society. But as much as the ladies of the frontier village of Shepherdstown must have longingly fondled the finery that filled Morrow's store—queensware dishes, china, ribbon, laces, cloth of all kinds including something called "cross bard negligie," sheeting and blanketing, women's scarlet coats, white and black silk mitts, combs, and books—they surely couldn't afford to buy much. Sales were slow, and neither Morrow nor the society ever saw much money from the deal.[11]

In the meantime, Rumsey succeeded in blocking Stanhope's English patent application and had received several more patents of his own. By now, Rumsey was a familiar figure at the patent office. During his time in England, he had received four patents covering more than twenty inventions relating to mills, steam engines, floating docks, and steamboats.[12]

During the summer of 1790, while Fitch was busy running passenger service on his Philadelphia steamboat, Rumsey was finally able to return to work on his steamboat. In June he got word that the patent act that had been passed in the United States in April, and he was not happy with its final form. "The law of Congress respecting patent rights almost amounts to an exclusion of my ever returning again to my own country," he complained in a letter to Levi Hollingsworth, then president of the Rumseian Society back in Philadelphia. "If judges should decide literally agreeable to the provision of that law, no patent in America can be worth holding." Rumsey apparently had interpreted the law to mean that variations to a device could be patented, opening the door to idea stealing.[13]

The following January, Rumsey had more bad news to relay to Morrow. "The persons I sold my patents to have failed, which at present has put a stop to all my affairs." Around this time, Parker's investors in the *Empress of China* scheme had tracked him down him in London. Rumsey found himself forced to pay Rogers and Parker's creditors, represented by a man named Maitland, about £300 for bills on which his name appeared. He "got paid with much difficulty otherwise I should have become bankrupt with them. I hope to keep the vessel and all the machinery from their concerns, if so she will keep me safe against debts that I owe on account of the concern to the amount of £800 sterling. The Society in America are mostly leeches and sharks, some of them has suppressed my letters to Barnes for some time back, in which was all Wynkoop's bills." The good news was that Rumsey had made progress on the steamboat. He told Morrow that it was "ready to move with a considerable degree of reputation, but must now be kept back from motives of policy."[14]

Sometime later that winter or in the early spring of 1791, Rumsey's worst nightmare came true. One of his creditors, whom he owed £68, sent the bailiffs after him. "[They] entered my room . . . and laid hold of me the moment I was out of bed, when ruffians of the most hideous appearance that were stationed at the door entered to conduct me to a prison." He bribed the ruffians to let him stay in his room until he could send for a friend, who was able to round up enough cash to make his bail—"by which I escaped a prison and a total overthrow of my affairs!"[15]

After that harrowing experience, Rumsey was back in hiding, his friends few and his creditors multiplying. By April, just as he was again fearing that a cell in a horrid London prison would be his next home, he was approached by two footmen in "loud livery" who ceremoniously presented him with an invitation. Confused, Rumsey nervously unfolded the paper and read it in disbelief. An Irish nobleman, the earl of Carhampton, was in London and wished to see him. The earl was looking to recruit a canal superintendent, and somehow, "from among the hundred thousand that profess to be engineers in this kingdom" as Rumsey put it, "he pitched upon me."[16] The two men met several times and soon made an agreement. Rumsey would be paid £10 sterling a day—a huge sum in those days, even for a consultant—for at least sixty days. Rumsey's head must have been spinning at this turn of good fortune.

The canal in question was undoubtedly the Royal Canal, which had been started in Dublin the previous year but had quickly run into problems that could be traced to inadequate surveying and planning.[17] Canal mania was beginning to spread throughout Britain and Ireland at the time, and this particular project had been spearheaded by an unhappy backer of the nearby and competing Grand Canal, which ran along a more southerly route from Dublin to the River Shannon. By 1789, more than £200,000 (the same figure quoted by Rumsey in his August letter to Morrow) had been raised for the project from shareholders and the Irish government. The workers had barely dug their way out of Dublin when they hit a mass of rocks in Closnilla. The shareholders—"meeting with difficulties that embarrassed and divided them," as Rumsey put it—sent Carhampton on a mission to find an engineer who could get them moving again.[18]

How Carhampton learned about Rumsey is a mystery. He must have talked to one of Rumsey's acquaintances in London who knew of his work for George Washington as the first superintendent of the Potomac Navigation Company. Rumsey's year-long stint five years earlier on the rocky parts of the river at Great Falls and Harpers Ferry was hardly canal work; mostly, Rumsey's workers tried to clear the river of boulders and other obstructions to navigation. But Rumsey had learned much about blasting rocks with gunpowder, which may have been exactly the experience Carhampton was seeking.

Rumsey left London for Dublin on May 6 and spent nearly three weeks getting there. He headed north to Liverpool on a roundabout route that took him past "all the principal canals of England," as he put it. These were probably the first completed canals he had ever seen. Certainly insecure about his limited knowledge of canal building, he must have wanted to learn as much as possible. Relieved to be out of the dirt and grime of London, he enjoyed touring the English countryside that spring, telling Morrow that "It would be a history to relate the particulars of this journey, which was attended with many scenes that were new and interesting, to a mind like mine."[19]

Once in Dublin, Rumsey was immediately called to appear before the board of the canal company. He wrote to his friend George West, who was suffering from tuberculosis and had returned home to Baltimore, "You know me too well my friend for it to be necessary to tell you the sensations I felt upon entering a room where undoubtedly was collected in about forty great men, a large proportion of all the science in the kingdom! Many were the questions I was asked and had to solve or answer." Rumsey was overwhelmed; the Potomac Navigation Company had not put him through the wringer the way these men did. He told Morrow the board had "worked up into the expectation (as some of the facetious told me) that I could create both clay and stone!" He worried that the hesitant and evasive way he answered their questions revealed his lack of confidence, and he feared they would see his inadequacies and send him on his way. But his considerable charm saved the day, and they put him to work. Rumsey never explained exactly what he did, other than to say that the job was completed earlier than planned and was a complete success after just forty days.[20]

Since his London creditors weren't expecting him back for at least another month, Rumsey took the opportunity to lay over in Liverpool for a while. Not eager to give up all of the £400 in his pocket, he instructed his new London agent, a man named Wakefield, to work up new terms with his creditors on a take-it-or-leave-it basis. Rumsey was determined to have enough money to work on his steamboat once he returned.

While waiting to hear back from Wakefield, he made the most of his time in Liverpool. Looking for ways to earn money, Rumsey decided to get back into the mill-building business, this time applying a few of his new inventions. He had never stopped working on ideas for mills. Of the several U.S. patents he received in 1791, two were for mill improvements. One of his new designs for a mill wheel called for the water to flow into a hollow wooden wheel that had two outlets on opposite sides. Water would enter through one outlet and be expelled out the other, causing the wheel to rotate, the action of which would turn a rotor attached to the wheel's hub.

That summer he set about building two test models to compare the hollow wheel idea with his own patented design for improvements to a type of mill known as Barker's. He found that the hollow wheel was more efficient and offered the considerable benefit of being able to work completely submerged in water, an advantage over conventional wheels when water levels rose. He called this his reaction-wheel mill, and he would receive an English patent for it the following year. He continued to work on its design, adding six internal wooden vanes to channel water through six outlets. A historian of technology has described this device as carrying "the seeds of the ideal water motor, capable of approaching 100 percent efficiency—the kind of motor to which the French were to give the name *turbine*."[21]

While in Liverpool, Rumsey came up with a plan to build a demonstration mill using the new wheel. The idea was for local millers to see the invention at work, then license its design from Rumsey. He sent his London creditors a letter that proposed a partial settling of his accounts by giving them a share in the planned mill. He warned that if they declined his offer, he would go ahead with a backup plan to form a partnership with several interested men in Liverpool, who had offered to pay him £300 up front. This sum would let him pay down his debts in London with enough left over to obtain a patent.

Rumsey's plan stirred his creditors into action. Fearing the inventor would never return, they quickly agreed to his terms. But when he came to London to close the deal, he was met with delays and "duplicity." He began to worry that if matters weren't settled quickly, the Liverpool mill, which was already under construction, would be completed and available for all to copy if his patent wasn't in place. Fortunately, an agreement was signed in time, "after it had undergone a thousand corrections . . . to relieve it of . . . ambiguity." Rumsey was gratified that the deal also called for "the haughty Mr. Maitland" to make good on the payment of a note Rogers owed Rumsey for £237.[22]

As eager as Rumsey was to return to steamboat work, he was not looking forward to living in London again. He wrote to Morrow about the gap he observed between the rich and poor: " . . . few of the innocent and substantial comforts of life exist near kings and courts, attendance and parade, such as continually go on in this great metropolis, where many are literally starving for bread, while others can't move a single yard without half a dozen servants to attend them, dressed in clothing more costly, by far, than our members of Congress! The wealth of this country is immense, many individuals spend their ten thousand a year without being half gratified for want of more!"[23]

His debts taken care of, Rumsey opened an office in the fall of 1791 and hired Wakefield to be his clerk. It was "at a place called Falean Starrs" on the Thames, near the southern end of the Blackfriars Bridge below the ruins of the Albion Mills. The location was ironic. Albion Mills—a striking Palladian brick structure—had been the largest and grandest steam-powered flour mill in the world, powered by a Boulton & Watt engine and closely managed by Boulton himself. This was the steam engine Thomas Jefferson saw while visiting London in 1786, shortly after it began operations; he later questioned Boulton about it.[24] The mill burned to the ground in March 1791, barely six months before Rumsey set up his office nearby. Arson was suspected. Even at this early stage in the Industrial Revolution, technology was viewed with considerable anger and distrust.

Rumsey hired at least a dozen men and put them to work in his shop, first to construct mill parts and later to work on the steamboat. The *Columbian Maid* was in sad shape after being abandoned for nearly nine months, and Rumsey recognized that she would have to be rebuilt. That winter, Barnes wrote to Charles Morrow about Rumsey's continuing troubles. Rumsey had told him he once thought he suffered more than his share of misfortune but had come to realize he lacked "a competent knowledge of mankind," which he wryly noted the "city of London is an excellent school to teach."[25]

Three months later, Rumsey wrote to Morrow that he longed to come home, but he knew he had to make one last try "in this land of exile. . . . [B]ut alas, my wearied patience must brace up to face another long year before I can even hope for that pleasure. . . . I have determined by that time to force my business here to an end of some kind or other. It might with certainty (should I live) be a prosperous end, had not all my connections here as well as my . . . patrons in America been leeches instead of liberal disinterested men! . . . Thus you see (as through all my life) I am obliged to live on hope, almost, alone." He advised Morrow to sell everything and buy a farm, for that is what he wanted to do: "It is in that way I hope to end my days!"[26]

In April 1792, Rumsey wrote to McMechen that business was once again at a standstill. His mills, which were to provide cash flow while he worked on his steamboat, had not come to much and his partners had seemingly abandoned him in the venture. How everything would end, "to use Barnes's expression, 'heaven only knows.'" At the end of June, he wrote that matters had worsened, with one of his mill partners in financial trouble and himself nearly broke again. "I am in good health and bad spirits."[27]

Somehow Rumsey managed to return to work on the steamboat in September, and by mid-November he moved it down the river to near Greenwich, away from the London crowds, where a test of the boiler proved successful. Rumsey was a few days away from making a full trial when another misfortune hit, on November 21. The workers had tied the steamboat to an anchored ship belonging to a charitable group called the Marine Society, with the captain's permission. But one of the society's members didn't like the idea and ordered the captain to set the steamboat loose. The captain untied the *Columbian Maid* during an ebb tide, and it drifted to shore and came to rest on its side. Rumsey was able to reach his boat within an hour of the event, only to find it had been punctured by the other boat's anchor and some surrounding wooden piles, leaving several large holes in the bottom. As a result, it took on water, so that when the tide returned it was too heavy to rise. Rumsey and his men worked furiously for six tides, until they could make repairs and get the boat afloat again.

On December 15, they tried the engine at the dock for the first time since the accident. "It worked with very good success. . . . It went forward against the tide and pulled hard to get from her moorings," according to Rumsey. His partner Rogers was on board and showed considerably more emotion, as Rumsey described in a letter to George West on December 18:

> Yesterday he [Rogers] met Mr. Parker on the stairs and gave him joy at what he was pleased to call an experiment, and wanted invitations sent off to the Duke of Clarence, Sir Joseph Banks, and others of the Royal Society, and, the people inflated with insanity talked of instantly putting engines upon ships of the navy against France (to protect the cause of despots against that of liberty, my friend); they talked of £50 to £60,000 as nothing to what they expected to receive. . . . It was with difficulty that I kept them from putting [a] wild scheme into execution immediately, and to wait until the experiment is actually made. Between ourselves, my friend, I have very little doubt of success, but it is really insulting to force me and the poverty of the business in this manner; it is now more than £200 in debt . . . I shall push forward the experiment for the sake of my character but have suffered too long to enter upon any new schemes that will run me into debt.[28]

Three days after he wrote this letter, Rumsey was dead. On the evening of December 20, 1792, he had presented a paper to the Society for Encouraging Arts, Sciences and Manufactures on the workings of a model of a waterwheel. The members present that night at the society's committee room in the grand Adelphi Terrace development recalled that Rumsey spoke mostly on hydrostatics and that he was well received. After the talk, Rumsey sat down at a desk to draft a resolution concerning his talk in the society's book. As he began to write, he raised his hand to his head and complained of a violent pain. He then collapsed. Two physicians were present and rushed him to a room in the Adelphi Hotel, where their efforts to help him failed. He died the following morning, at the age of forty-nine. His friend Richard Claiborne stayed with him through the night. In a letter to the Rumseian Society, Claiborne wrote that "Mr. Rumsey had his reason at intervals, tho' speechless and helpless, and a few minutes before his death whispered, "[I] must die!"[29]

The sad news was relayed to Rumsey's friend George West in Baltimore by Rumsey's clerk, Wakefield. "Unhappy the task that sincere friendship here imposes—our friend, alas! is no more."[30] He told West that Rumsey was buried at St. Margaret's Church, adjacent to Westminster Abbey, on December 23, and that he had sealed his papers to await instructions from Rumsey's family.

Thomas Jefferson got word of the tragedy in a letter from John Brown Cutting, an American physician living in London. He noted that an autopsy had been performed to determine the cause of death. Rumsey's skull was opened, and "What I had before conjectured now became confirmed—that overplied with energies of thinking some of the vessels of the brain were fairly worn out. Accordingly there had happened a rupture of one of them—which was manifestly the immediate cause of his death." Cutting told Jefferson that the long-awaited public trial of the *Columbian Maid* was to have taken place the day Rumsey died.[31]

Some time in January or February 1793, Rumsey's steamboat was tested on the Thames. According to an account published a few months later in a London magazine, it made four knots ~~an hour~~; no distance was specified.[32] After that, the *Columbian Maid* went silent.

Thanks to the slow mails, Rumsey at least had been spared the bad news that his brother-in-law Charles Morrow had died a month before him. His other brother-in-law, Joseph Barnes, was left to sort out Rumsey's affairs in England and America, with some help from Jefferson, who in May 1793 wrote letters of introduction for Barnes to take to Jean-Baptiste Le Roy, the French scientist; Thomas Pinckney, then U.S. ambassador to England; and Gouverneur Morris, U.S. ambassador to France.[33]

A letter from Matthew Boulton to James Watt in May of the following year mentions that an American, probably Barnes, was in England to "take possession of Rumsey's property, which consists of a patent or two and some debts." Boulton said that he had met with Barnes that morning and that Barnes had asked him for a written account of any experiments he and Watt may have conducted in steam navigation, "All of which I evaded," Boulton remarked, "and am to see him and Rumsey's partner [Parker] next week to talk over the matter."[34] In April 1795, Barnes settled Rumsey's debt with Parker in the amount of £86. Rumsey's wife and children back in Virginia received nothing.[35]

Daniel Parker wasted no time in making good on his failed investment in Rumsey. Although the details are not known, another former business partner of Parker's, Joel Barlow—the American whom Rumsey had once considered using as an agent—took Rumsey's English patent specifications for his tubular boiler and in 1793 applied for a French patent of importation.[36] Barlow's action was purely speculative but oddly coincidental. Four years later, he would meet Robert Fulton in Paris and help Fulton with his first steamboat experiments. A tubular boiler would be part of the plan.

Perhaps not coincidentally, less than six months after Rumsey's death, Robert Fulton, who was still pursuing a lackluster art career in England, began thinking seriously about steamboats. An undocumented letter from Rumsey to George West is said to reveal that Fulton had been friendly with Rumsey while both were living in London, according to Rumsey's biographer.[37] The acquaintance was certainly possible; both West and Fulton were around that time students of the American artist (and a founder of the Royal Academy of Art) Benjamin West in London.

Interest in steamboats dwindled in Europe in the final years of the eighteenth century. In Scotland, Patrick Miller, a retired Edinburgh banker, built a paddlewheel steamboat in 1788 with the help of the mining engineer William Symington. The trial, carried out on a small lake on Miller's estate, was successful but not encouraging. Miller put up more money to build a larger boat; it was tested the following year and performed poorly.[38] In 1790,

James Rumsey, portrait by Benjamin West, c. 1790.
West Virginia State Archives

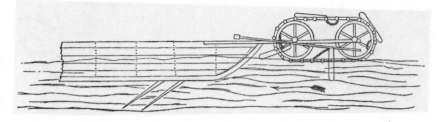

James Rumsey's mechanical poleboat

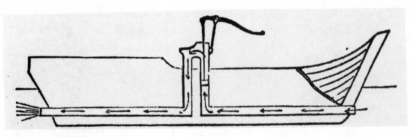

Benjamin Franklin's plan for a water-jet-propelled boat, 1785
From Thompson Westcott, *Life of John Fitch,* 1878

Model, James Rumsey's 1787 Shepherdstown steamboat
Courtesy of the Smithsonian Institution, NMAH/Transportation

A

SHORT

TREATISE

ON THE APPLICATION OF

S T E A M,

WHEREBY IS CLEARLY SHEWN,

FROM

ACTUAL EXPERIMENTS,

THAT

S T E A M

MAY BE APPLIED TO PROPEL

BOATS OR *VESSELS*

OF ANY BURTHEN AGAINST RAPID CURRENTS WITH GREAT VELOCITY.

The same Principles are also introduced with Effect, by a Machine of a simple and cheap Construction, for the Purpose of raising Water sufficient for the working of

GRIST-MILLS, SAW-MILLS, &c.

AND *for* WATERING MEADOWS *and* OTHER PURPOSES OF AGRICULTURE.

By JAMES RUMSEY,

OF BERKELEY COUNTY, *Virginia.*

PHILADELPHIA,

PRINTED BY JOSEPH JAMES: CHESNUT-STREET.

M,DCC,LXXXVIII.

Cover page, Rumsey's pamphlet, 1788
Courtesy of the Smithsonian Institution, NMAH/Transportation

John Fitch, fresco by Constantino Brumidi, in the U.S. Capitol
Architect of the Capitol

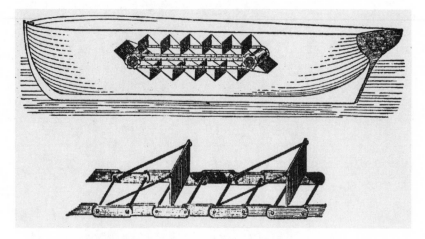

John Fitch's first steamboat design featuring an endless chain of paddles
From Thompson Westcott, *Life of John Fitch,* 1878

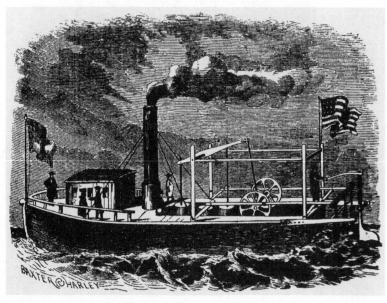

John Fitch's stern paddle Philadelphia steamboat, 1788–90
From Thompson Westcott, *Life of John Fitch,* 1878

Designed by Reigart.

FITCH'S STEAMBOAT.

On the Delaware River, opposite Philadelphia.

L. N. Rosenthal's lith. Philad.

1562

John Fitch's 1787 Philadelphia steamboat, featuring his crank and paddle design, from a painting by John Franklin Reigart. (The artist was wrong on two counts: this boat was named *Perseverance*, and it first operated in 1787.)
Library of Congress, Prints and Photographs Division, LC-USZ62–1362

William Thornton, portrait by George B. Matthews, 1930, after Gilbert Stuart
Architect of the Capitol

John Stevens
Stevens Institute of Technology, Library Special Collections, Castle Point on
Hudson, Hoboken, New Jersey

Oliver Evans

An artist's depiction of the *Orukter Amphibilos,* Evans's 1805 steam vehicle/river dredger, Philadelphia
Harper's New Monthly Magazine, October 1868

Washington, D.C., in 1800, by Conrad Malte-Brun
Library of Congress, Prints and Photographs Division, LC-US-Z62–102054

Robert Fulton, by Benjamin West, engraved by George Parker
Library of Congress, Prints and Photographs Division, LC-US-Z62–106813

(opposite, bottom) Fulton's *North River*
steamboat, 1807 (aka "Clermont")
Library of Congress, Prints and Photographs
Division, LC-US-Z62–1342

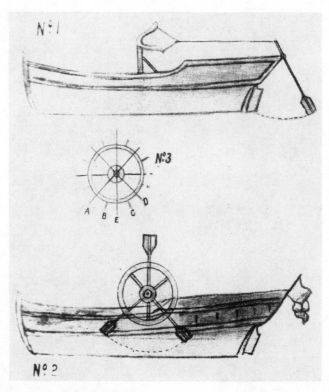

Fulton's first steamboat plan, 1793
From Alice Crary Sutcliffe, Robert Fulton and the "Clermont," 1909

Benjamin Latrobe, by George B. Matthews
Architect of the Capitol

The 1911 replica of Fulton's 1811 steamboat *New Orleans,* the first steamboat on the
Ohio and Mississippi Rivers
Thomas E. Tappan Collection, Special Collections, University of Arkansas Libraries,
Fayetteville, Arkansas

John Stevens's *Phoenix* steamboat, 1809
Stevens Institute of Technology, Library Special Collections, Castle Point on Hudson,
Hoboken, New Jersey

John Fitch's model for a steam locomotive, c. 1796
Ohio Historical Society

Miller wrote to Boulton & Watt proposing they work together on a steamboat. Watt replied with a discouraging letter, saying that the company had no interest in any new ventures, particularly those having to do with steamboats. Miller dropped the idea after that.

Back in America, a man in Rhode Island named Elijah Ormsbee, who had heard that someone was running a steamboat in Philadelphia, partnered with David Wilkinson, a Pawtucket iron founder, to build a steam engine for use on a boat. Ormsbee was said to be inspired by the Newcomen steam engine that was still pumping water out of the iron mine near Cranston. Ormsbee tried out his engine on a borrowed boat, using "goose-foot" paddles (he rejected paddlewheels) and a Newcomen-type engine, in a cove of Narragansett Bay in the fall of 1792. It achieved a speed of a little more than three miles an hour. He tinkered with steamboat ideas for several more years but never built anything but models.[39]

In 1793, John Stevens was still hoping to build steamboats. He hired John Hall, who had worked as an engineer for Thomas Paine and John Fitch, to build an engine in New Jersey. But Stevens was forever standing over Hall, offering unsolicited ideas and unwanted instructions. Hall turned to drink, which led to his firing a few months later.[40] Unable to find another qualified mechanic, Stevens set the idea aside.

Without Rumsey, the steamboat *Columbian Maid* disappeared from the banks of the Thames in early 1793, her fate unknown. Around that time, John Fitch was on his way to France, eager to make one more try.

TEN

MOTHER CLAY

He that studies and writes on the improvements of the arts and sciences labours to benefit generations unborn, for it is improbable that his contemporaries will pay any attention to him.

—Oliver Evans

John Fitch, who suffered on ships, was grateful to be on land after a miserable two months in steerage across a wintry Atlantic. His first steps in a foreign country were made, a little shakily, in the busy port city of La Rochelle in western France. It was April 10, 1793. As Fitch's luck would have it, his arrival coincided with the start of the Reign of Terror.

King Louis XVI had been beheaded in late January, but the news had not reached America before Fitch left. The incident further divided France. The previous September, the French Convention had abolished the monarchy and formed a republic. By March, the royalists in the Vendée, the coastal region between La Rochelle and the Loire estuary, began a counterrevolution. Fitch stepped off the ship and found himself in the middle of a civil war.

As soon as he landed, he wrote to Vail in Lorient and asked for instructions on how to proceed. While he waited for a reply, Fitch explored the town and surrounding countryside, trying his best to fit in. It was the custom at that time for Frenchmen to wear the tricolor cockade, a ribbon rosette, on their hats to show support for the new republic. He wrote to Thornton: "When I first came to this town I had many serious scruples about putting the national cockade into my hat for fear it would have the appearance of denying my own nation, which heaven knows I never will do but willing to show my patriotism . . . with much reluctance tucked one under my hatband."

To soothe his patriotic guilt, he told Thornton, he had unwittingly committed a political faux pas: "[I] shortly after drew the American colours on paper with the stars and stripes and placed it by its side but was advised and I believe prudently to take it out, which I did." Fitch was overwhelmed when, a few days later, a "noble, generous Frenchman" raised the American flag on the flagpole in the public square, presumably in honor of the town's visitor.[1]

Fitch walked into the same square later that month to find that a stage had been built and a strange machine called a guillotine was being moved into place. To his dismay, the American flag was still aloft on the mast, as if to condone the horrific events to come. His letter to Thornton continued, "I saw the execution this afternoon and however shocking it might be to my feelings it did not prevent me from noting that a large part of the spectators were composed of women. It gave me pain to see the most delicate part of the human species so forward and unconcerned at so dismal a spectacle as they generally appeared to be. But what gave me the most pain of all as soon as the executioner showed his head when severed from the body was the whozzas of the people. . . . [T]o see the same square stained with human blood that is adorned with American colours I must confess I am not altogether reconciled to the sight."

A few days later, after hearing from Vail, Fitch left for Nantes, where Vail had arranged for Fitch to get castings made for a steam engine. Again, Fitch's timing was terrible. In another letter to Thornton, Fitch wrote, "I went to Nantes . . . and soon after my arrival the furnace became a frontier from the insurrection in that part of the country and shortly after the town was besieged, when I made my escape out it."[2]

Nantes had been suffering, like all French ports, from naval blockades imposed by the several countries with which France was at war. Worse, the civil war was intensifying. When the guillotines could not chop off heads fast enough, Robespierre ordered the mass drownings to begin. Thousands of "enemies of the republic"—women, children, nobility, and priests, sometimes tied together—were put on barges that were sunk in the Loire.

Fitch made the ninety-mile trip up the coast to Lorient without incident, arriving on May 18. He settled down in Vail's home and waited for conditions to improve. Fitch wrote to Thornton a few days later: "We had news day before yesterday of a large body of aristocrats being defeated somewhere in the county, I believe between Nantes and Tours, where they lost fifteen or sixteen hundred men and fourteen pieces of artillery. . . . I could have wished instead of their being killed that they had all been landed safely in America. I think that we could manage them by our mild gentle laws so as to make them good citizens of our commonwealth without putting so many to death as they do."[3]

Thornton had given Fitch a letter to take to their French friend, Jacques Brissot de Warville, who had traveled in America in the latter half of 1788 and had seen Fitch's steamboat in Philadelphia. Brissot was now the leader of the republican Girondist party in France, with support from Thomas Paine, another Fitch and Thornton friend. Fitch never made it to Paris to deliver the letter, but it was just as well. By early June 1793, the Girondists, a group of moderate republicans, had been forced out of their seats in the National Convention by the Jacobins and the sans-culottes. Brissot and many others were arrested later that month, and they were tried and beheaded that fall.[4]

By August, conditions were bad everywhere. Fitch lost all hope of building a boat and engine in France. He wrote that month to Thornton, "I need not inform you that it is impossible to go on with the steam boat in this convulsed situation of the empire of France, for your papers will undoubtedly satisfy you that I cannot."[5] Had Fitch come a year earlier, instead of trying to launch *Perseverance II,* steamboat history might have taken a different turn.

Vail, though, was not ready to give up. He paid Fitch's passage to England and gave him a letter of credit to have a steam engine built, presumably by Boulton & Watt. The fact that Britain and France had declared war on each other earlier that year—a war that would last until 1815—did not seem to Vail to be a problem. He also gave Fitch a letter of introduction to Joshua Johnson, the U.S. consul in London, whom Vail had known in Nantes a few years earlier and who he hoped would offer help.[6] Fitch sailed from Lorient that fall and arrived in London in mid-November 1793.

Fitch never recorded his impressions of London, but he must have been taken aback. In 1793 it was the world's largest city, with nearly a million people living in crowded and largely filthy conditions. Fitch found Johnson at his office, explained his contract with Vail and their current problems, and asked for advice in arranging for an export permit for a Boulton & Watt steam engine. Johnson looked at Fitch aghast. Didn't Mr. Fitch and Mr. Vail realize that Britain and France were at war? Export permits were a thing of the past. Even if a permit were possible, Britain was not about to help France gain a foothold in some new technology with its own machinery. Defeated, Fitch pulled out the letter of credit from Vail and handed it to Johnson. Could he lend Fitch the cost of passage back to Lorient? Again Johnson laid out the bad news: It was considered "high treason to negotiate French bills of exchange." Johnson did offer to use Vail's letter to advance the cost of passage to America, but Fitch proudly declined. He could not in good conscience use Vail's money to quit and go home. Fitch thanked Johnson and left.[7]

It is unclear why Fitch, during his time in London, did not apply for a British patent. He may have thought that Rumsey's patent there would preclude him from such a move, although the boats and engines were quite different in design. Or perhaps he had heard about the Boulton & Watt caveat on file at the patent office regarding the use of steam engines in boats. Quite possibly it was an issue of money—the £100 application fee certainly discouraged all but the most serious or wealthy.

Fitch was staying with another American, Robert Leslie, a fellow inventor friend of his from Philadelphia, in Leslie's home off Lower Thames Street, near London Bridge. Leslie had moved to London the year before to set up shop as a clock- and watchmaker with his partner, Isaac Price.[8] Leslie held five U.S. patents for various mechanical devices, but he was best known for his improvements to timepieces.[9] It seems likely Leslie would have told him about Rumsey's death and possibly the fate of Rumsey's steamboat, which had made its only trial on the Thames just ten months earlier.

Thanks to Leslie's generosity, Fitch had a roof over his head, but he badly needed money for food, clothes, and passage home that spring. In a scheme reminiscent of his Northwest Territory map a decade earlier, Fitch decided to publish and sell a navigation chart he called the "Ready Reckoner." On his Atlantic crossing, he had observed how the ship's captain kept track of location by using a complicated system of instruments and calculations. With little else to occupy his time, Fitch devised a one-page chart that served the same purpose. He sent a copy to Thornton from Lorient that August, asking him to take out a patent on it. While staying with Vail, he pounded out a sheet of copper and did his own engraving, as he had done for his map.[10]

Fitch took his copper plate to Gilbert and Gilkerson, London printers, and instructed them to print it along with an accompanying pamphlet of instructions, in which he offered an early version of a help desk: Any purchaser "who does not readily understand it from the plate and the description can call on the author between 10 and 12 o'clock, Sundays excepted, till they are perfect masters of it at half a guinea each."[11]

Around Christmas, he took the printed copies down to the Thames wharves and tried to sell them to ship captains. To his complete surprise, he was met with scorn and derision. Why, one captain after another asked him, would they want to put themselves out of a job? At least Thornton offered encouraging words about the project, although it wasn't until February that he wrote to Fitch again. "I cannot help but approve your design, for in my second voyage on the Atlantic I made a chart of the same kind, and applied it to the very same use, but mine was neither so extensive or comprehensive. It is in my opinion an ingenious method, for which you have a right to much credit. Mine was exhibited, indeed at the time I invented it about seven years ago, to Capt. McEvers of this place, when at sea, and he said it was similar to a French quarter card." He added that he would take care of getting Fitch a patent for it.[12]

Thornton in the same letter told Fitch that he was also writing to his old friend in London, the prominent Quaker physician John Coakley Lettsom, "one of the best of men," on Fitch's account. " . . . I know he will be glad to see you, therefore wait on him when you receive this. Do not fight the Algerines before him. You know what I mean. Regulate your genius." This was Thornton's way of advising Fitch to put a lid on his usual ranting about how steamboats could offer navies a huge advantage in fighting the Barbary pirates. Lettsom, who had founded the London Medical Society, may have recalled this wasn't the first time he had been asked to help out an American steamboat inventor; five years earlier, the equally prominent Philadelphia physician, Benjamin Rush, wrote a letter of introduction to Lettsom for James Rumsey.[13]

Thornton also mentioned the "dreadful havoc" caused by the yellow fever that year, saying, "Mr. Wells had it very ill, but got better. The rest of our friends escaped." It seems Fitch's one and only bit of luck in recent years was that he left Philadelphia just months before the yellow fever epidemic of 1793, which killed one-tenth of the population, about four thousand people, and closed down the federal government for several months. Thornton himself had contracted a mild case.[14] He ended his letter with this stern admonition: "Let me advise you to get no steam engine made, *except* by *Watt and Boulton*, and *with a copper boiler, without any wood round it*, and *very strong copper.*"[15] (Emphasis his.) By the time Fitch would have received this letter, certainly no earlier than mid-March or April, Thornton's advice was irrelevant. Fitch had probably already left London. Broke and with no way to return to France, Fitch had given up. (Lettsom later wrote to Thornton that he never heard from Fitch.)

Sometime that spring, Fitch boarded a ship to America, contracting with the captain to pay back the cost of his journey—twelve and a half guineas—by working on the Boston docks upon his arrival. The ship reached Boston on June 6. Fitch took a room at the Wing's Inn near the waterfront and worked as a dock laborer. The experience was therapeutic. He wrote to Thornton later that summer that "the happiest days of my life is since I came to Boston. . . . My labor is an amusement and affords me a moderate sustenance, and my accommodations are modest and agreeable, and I live retired and unknown, and go every day in the fore and afternoon with my fellow journeymen with the greatest freedom and pleasure to water."[16]

Fitch hoped to pay back the captain quickly, and to that end he wrote to Thornton the day after he arrived to make him an offer. He proposed to sell him half the interest in the Ready Reckoner for the amount of his debt, or, if Thornton wasn't interested, to strike a similar deal on his behalf with John Nicholson. It seems neither man accepted.[17] Fitch received a letter from Thornton later that month, but a copy has not been found. Fitch replied with the good news that a Boston mathematician named Osgood Carleton thought so highly of the Ready Reckoner that he and Fitch made a deal: In exchange for one-half the proceeds of its sales in New Hampshire, Massachusetts, and Rhode Island, Carlton would pay the cost of printing 300 copies. Nothing more is known about this scheme; it appears to have fizzled.

Fitch had one card left to play. He told Thornton that while he was living in London he came up with a new way to turn millwork, adapting his crank and paddle device, to watermills. "I am now about making a model of the same, which will be completed in a few days, to ascertain its force and utility." He included a drawing of his plan in the letter and asked Thornton to apply for a U.S. patent for him. He hoped to sell his patent rights once the device proved successful. Thornton's notes show he entered a patent application with the secretary of state on August 4, but there is no record that Fitch ever received a patent. By late July he was close to forming a partnership with two men who were interested in his device for a woolen mill in Byfield, but Fitch withdrew at the last minute, not liking "their nearness in bargaining." He asked Thornton to make sure one of the men didn't show up in Philadelphia seeking a patent on the same idea.[18]

After this, Fitch never heard from Thornton again. For years after, he feared his dear friend had abandoned him. He wrote to him in late September asking why he had not written; Fitch's own insecurity led him to think that Thornton may have distanced himself from the inventor for political reasons: "But Sir, however unpopular it may be at this time to do one justice, I am persuaded the time will come when it will not be thought disagreeable to be called my friend."[19] Fitch may have known that Thornton had been named by President Washington as one of three commissioners of the fledgling city of Washington. Besides the demands of his job, Thornton had another reason for tardiness in letter writing. He had been ill that summer with a "bilious fever."[20] If Thornton did write back, his letter again may have arrived too late. By January 1795, Fitch had left Boston, destitute once more. It is not known if he quit or was fired from his dockworker's job, or if he ever paid back the cost of his Atlantic passage.

According to two of his biographers, Fitch's next stop was his hometown of Windsor, Connecticut, after an absence of twenty-five years. Perhaps he sought to reconcile, if not with his wife, at least with his children. If he made the attempt, it does not appear he was successful. These biographers note he stayed in Windsor for about a year, living with his older sister Sarah and her husband, Timothy King. But one of Fitch's letters to John Stevens that summer mentions that he was staying in Sharon. Although one biographer states this was Sharon, New York, it seems more likely that it was Sharon, Connecticut, which was Timothy King's birthplace. It is possible that Fitch's sister and her husband were living there, not Windsor, at the time. From this home base, and probably with some financial help from King, Fitch was able to travel a bit.[21]

In April, he visited John Stevens at his home in Hoboken, New Jersey, trying to interest him in the horseboat idea. "Mr. Fitch called on me a few days ago," Stevens wrote to John Nicholson later that month, "with a proposition of being concerned with him in erecting . . . a horse boat. He at the same time informed me that he had conveyed to you the greater part of the interest in the patent he had obtained therfor." Stevens explained that he ran a ferry service from South Amboy to New York City and was intrigued by the horseboat idea. He asked Nicholson to relate his experience with Fitch in building a horseboat and if Nicholson would be interested in building such a boat in New Jersey.

Stevens also wrote to Fitch's friend and former Steamboat Company shareholder Benjamin Say, inquiring about the boat's defects and advantages. Dr. Say replied that he had seen the boat the previous fall—"a double boat with a platform extended across between them"—but didn't know enough about it to help Stevens.[22] Fitch continued to bombard Stevens with letters about horseboats through the summer: "[T]his, sir, is as impossible of failure as that four horses would fail of drawing a wagon." Fitch badly wanted Stevens to put him in charge of building the boat, in return for "a moderate sustenance."[23] Fitch hoped to have a boat ready by the following spring, and to that end he wrote to Stevens a few more times. In one letter he said that he wanted to "make a conveyance of one-half the patent right on the North and Raritan rivers to the person who will take me by the hand so far as to get one boat agoing complete." Fitch was by then into the eighth year of his fourteen-year monopolies to run steamboats on New York and New Jersey waters. But Stevens apparently decided against the idea, passing up an opportunity that would have helped him greatly a few years later.

By August, Fitch gave up on Stevens and continued on to Philadelphia. It seems certain Fitch found his boat in ruins at the docks. He decided to sell its parts at auction to pay off the Steamboat Company and bring in a little income. An ad in the *Aurora and General Advertiser* for August 18, 1795, signaled the end of Fitch's dream:

A STEAM ENGINE.

On Wednesday, the 24th inst., will be sold by Public Vendue [*sic*], on Smith's wharf, between Race and Vine streets, a sixteen inch cilinder steam engine, with machinery appertaining thereto. The terms of the sale will be cash, and the sale to commence at ten o'clock in the morning. Composing the same there are, viz.:

A COPPER BOILER, with 2 large pipes, cocks, &c.
A 16 INCH CAST IRON CILINDER, STEAM CREST, PISTON, ROD, CHAIN, &C.
A LEADEN SINK PIPE & BRASS VALVE.
A LEADEN PIPE AND COCK, for supplying the piston.
One do. for the waste water.
One LEAD CILINDER CUP.
A LEAD PUMP for injection water.
2 CISTERNS.
1 large fly wheel (cast iron) and AXLE thereof.
2 TWO FEET CAST IRON WHEELS, handy for steam and injection.
A FURNACE DOOR and GRATING.
A 9 or 10 feet LEVER or BEAM.

PUMP RODS and boxes for do.
A SMOKE pipe, and sundry other apparatus, &c.

EDWARD POLE,
Auctioneer.[24]

Fitch visited several old friends and supporters while in Philadelphia, with the possible exception of Henry Voigt, who by then was a bigwig at the U.S. Mint. This visit seems to have been the timeframe referred to in an 1814 deposition by the inventor Oliver Evans, who said that Fitch came to see him at his house, declaring "his intention to form a steamboat company to establish steamboats on the western waters."[25]

Fitch missed by months meeting a fellow steamboat schemer in Philadelphia, a fleeting and mysterious mechanic named Griffin Greene.[26] Greene, then in his late forties, was a forge master from Rhode Island who settled in Ohio in 1789. There, in partnership with iron works owner Elijah Backus, he built a floating grist mill on the Ohio River sometime after 1790. In 1795, the Treaty of Greenville was signed, specifying the rights of Indians and settlers in the region, an event that was supposed to make river travel safer.

Greene and Backus had heard about the steamboat wars of Fitch and Rumsey. Seeing that the original inventors had quit the task, the two men formed a partnership to build a steamboat. Details are lacking, but by April 1796 construction of the engine was well underway in Philadelphia, and Backus received a U.S. patent for it in late May. A successful trial with a model boat in mid-May led Greene and Backus to contract with a shipbuilder named John Patterson to construct a fifty-foot-long steamboat. Starting with a $300 down payment to Patterson, the costs mounted quickly. By September, debts of more than $10,000 had been rung up, an enormous sum for those days. Bankruptcy followed, and Greene's steamboat never left the dock.[27]

Fitch, meanwhile, made plans to return to Kentucky. He hoped to make one last try at a steamboat by forming a company there or in Pittsburgh. He planned to pay for the venture by selling the 1,300 acres he owned near Bardstown, the land he had successfully gained title to after his 1781 surveying trip to the area.[28] He arrived sometime in early 1797 to find an area quite different from the one he left fifteen years earlier. Kentucky, by then the fifteenth state, was becoming civilized. Lands had been cleared for farms, and towns had grown out of the forests; in fact, the village of Bardstown had grown up near his lands. His first stop was a tavern owned by Alexander McCowan. He made McCowan a deal. He explained that he expected to die soon, and if McCowan would provide him with room, board, and a pint of whiskey a day, he would give him 150 acres of prime neighboring farmland. McCowan later said that he thought Fitch looked healthy enough but, seeing the potential for profit, he agreed.

Fitch set out to check on his properties and was upset to find that six different families had built houses and barns on the lands and had cleared the fields. With his Virginia deeds in hand, he went to see the local attorney, a young man named John Rowan. Rowan agreed Fitch had a case and began filing lawsuits against the squatters.[29] It seems Rowan was eventually successful to some degree, although the process was long and difficult. He apparently told Fitch that the lands were worth as much as $13,000, a considerable fortune at the time.[30]

To occupy himself while his legal battles dragged on, Fitch borrowed space in the local blacksmith's shop and built two models. One was a new kind of steamboat, and the other was something else altogether. About two feet long, it was odd looking, consisting mostly of a steam engine mounted on a wood base, with brass workings and four flanged wheels resting on a track. It most definitely was not a boat. Any papers he may have left behind describing it have been lost.

Fitch never stopped talking about steamboats and steam engines. His first biographer, Thompson Westcott, corresponded in 1855 with Robert Wickliffe of Lexington, an attorney and state legislator who was a young man in Bardstown during Fitch's time there. Wickliffe had distinct memories of Fitch. He wrote to Westcott, "Those who were intimate with him assure me of their belief that Fitch's profound mortification in being compelled to abandon his steamboat discoveries, and the new difficulties and legal controversies about his land titles, broke down his spirits and disgusted him with life. . . . McCowan . . . informed me that it was the constant burden of his conversation, when free from intoxication, that he should descend to his grave poor and penniless, but should leave in his discoveries a legacy to his country that would make her rich."[31]

Referring to the arrangement Fitch had set up with McCowan, Wickliffe wrote that one day Fitch said to McCowan, "'I am not getting off [drinking himself to death] fast enough; you must add another pint, and here is your bond for another one hundred and fifty acres of land.' Both of these bonds McCowan showed me, and got me to read them."[32]

Fitch had been in poor health for a year or two. He wrote one last sad letter to Thornton on February 1, 1798. Fitch had not written to his old friend for more than three years, probably because he never received replies to his letters of late 1794. But now seemed to be the time to wrap up loose ends. He began,

> Worthy Sir, I am going fast to my mother clay. Yesterday I executed the last will I ever expect to make. My will is as followeth—I have given to Mrs. [sic] Eliza Rowan, an infant daughter of Jno. Rowan Esqre., attorney at law, the one-thirteenth part of all my freehold estate—all the remainder both real and personal I have given to you, to Mr. Rowan, Mrs. [sic] Eliza Vail, daughter of Aaron Vail Esqre. of L'Orient, and to my two children in equal proportions share and share alike. All I request of Mrs. Eliza Vail for the legacy I have left her, as I shall never see my two Eliza's meet, that when she is big enough, be her in what quarter of the globe she may, to hold a constant correspondence with Mrs. Eliza Rowan. . . . When I was lying on the downs I wrote a long circumstantial letter to Mr. Vail in hopes that I should come across some French vessel at sea that if they would not take me again to France that I could at least forward the letter. Being unfortunate in this till I arrived in Boston I suppressed it, for I could not endure the thought of hurting his feelings by my misfortunes and the hard usage I met with in the steerage . . . or to inform him that I was brought to the alternative of working in Boston as a journeyman. I did not mean to write to either you or him till my business in this country should be fully settled but finding myself grow weak so fast am fearful I shall not see the issue of it.

Fitch then relates the story of his unsuccessful attempts to gain Johnson's help in London and the circumstances of his sea passage back to America. He asks Thornton to convey the contents of his letter to Vail and to encourage Vail, should he come to America, to settle in Kentucky. He closes with this: "My worthy friend I have many more things to inform you and Mr. Vail but being fatigued shall only say that I am and shall die a friend to both of you." He signed the letter and added a poignant afterthought: "P.S. If possible let me receive one more letter from you."[33] If Fitch could have known what Aaron Vail would do with his steamboat plans three years later, he would not have offered such warm thoughts.

There is no record that Thornton responded. Fitch made a new will on June 20, with several noteworthy changes to the will he described to Thornton. Added were William Rowan, father of John, and James Nourse, a Bardstown friend. Left out from the original version were the infant Eliza Rowan and Fitch's two children. Fitch did not explain the reason for omitting his children from his will, but one guess is that he tried to contact them after writing the first will and, receiving no response, left them out. It is not clear why he dropped Eliza Rowan; one account notes that the child had died in the meantime, but that was not the case, according to genealogical records. His fondness for his "two Elizas" surely reflected his regret that he never knew his own daughter.

Fitch's biographers have all stated that Fitch committed suicide shortly after making this new will. Fitch no doubt was deeply depressed, but a Fitch descendant believed he died from either heart or liver failure. Roscoe Conkling Fitch traveled to Bardstown in 1905 to try to learn the truth, albeit more than a century after the fact. "I talked to everybody who had any information at all. As I got it from them, and as I believe it to stand for the truth . . . Fitch went out from the inn, while ill, to a spring down the bank and that his heart failed him when he tried to return. It was night. He was found the next morning early but he died that day."[34] The likely date of his death was July 2, 1798. He was buried in a casket made of cherry wood in the Bardstown public cemetery, with little ceremony and no headstone.

Thirteen years later, Fitch's bones would be rattled by the first of three magnitude 8 earthquakes unleashed by the New Madrid fault during the winter of 1811–12. Just a few days before it struck, the first steamboat on the Ohio River had passed within twenty miles of Fitch's grave, on its way to New Orleans.

ELEVEN

THE FRENCH CONNECTION

Blast his belly! He stole my patent!

—*Samuel Morey*

John Fitch may have been going fast to his mother clay in the spring of 1798, but had he known what people were saying about him in New York, he might have rallied for one last fight.

Robert R. Livingston—better known as Chancellor Livingston (he was chancellor of New York State for many years)[1]—was a wealthy lawyer and public servant who had served on the committee to draft the Declaration of Independence and in the Continental Congress. At the moment he was obsessed with the idea that fortunes could be made operating steamboats on the Hudson River. Long an important waterway, the Hudson conveniently flowed past his estate, Clermont, about ninety miles north of New York City. He saw that it had much more commercial potential than the Potomac or the Delaware, with the added advantage of being wide, straight, and relatively free of obstructions. Livingston could see the day when the Hudson would link New York to the West via a system of canals across the western part of the state to the Great Lakes. In fact, canal building was already underway in New York by the mid-1790s, although plans for the Erie Canal were still several years away.

Livingston had been mulling over boat ideas for a couple of years. He first thought of building a horseboat along the lines of Fitch's plan, in which the horses would turn a horizontal wheel. But unlike Fitch's boat, the wheel would be mounted in a box under the keel, which would then turn and spew out water, propelling the boat forward—a variation on the jet-propulsion approach advocated by Franklin and tried by Rumsey. He talked the idea over with his brother-in-law, John Stevens, who had rejected Fitch's horseboat deal three years earlier. Stevens advised him that steam would be a better bet than horses.[2]

It seems likely that Stevens also advised Livingston to get the state legislature to repeal the act giving exclusive rights to navigate steamboats to John Fitch back in 1787. Stevens suggested this because he had tried and failed to obtain those rights himself. Even though Fitch's monopoly still had three years to run, Livingston was impatient to get started. He pressed his friends in the state legislature and convinced them that Fitch was either dead or gone missing. They repealed Fitch's award on March 27, 1798, and then gave Livingston a similar monopoly—except that this one was for twenty years, if he could move a boat at the speed of four miles per hour between New York City and Albany within a year. The legislators had no problem granting such a long term. Laughing among themselves, they called the act the "hot water bill" and were sure that Livingston, known to be a tad eccentric, would never get a steamboat to work.[3]

The event that sparked Livingston's interest in steamboats happened two years earlier. In the spring of 1796, a thirty-five-year-old New Hampshire inventor named Samuel Morey came to New York looking for financial support. He invited Livingston for a ride on his steamboat, which was propelled by a stern paddlewheel. The chancellor brought along his brother Edward, a cousin, and Stevens. They "went with me on the boat from the ferry as far as Greenwich and back, and they expressed great satisfaction at her performance and with the engine," Morey later wrote. The following year, he tried another steamboat in New York, this one with two side-mounted paddlewheels, which he thought worked even better.[4]

Morey had been developing his steamboat ideas since 1791, quite independently of the other steamboat inventors. In 1793, he made a public trial of a steamboat on the Connecticut River near his home in Orford, New Hampshire. This boat was a one-person dugout log canoe with a paddlewheel at the bow that made an estimated speed of five miles an hour upstream. The boiler and engine took up most of the space, with just enough room left over for Morey and a supply of firewood. Earlier that year he had received a U.S. patent for a steam-powered spit, helpful in the unattended roasting of meats. Two years later he received the first of his two patents for steam engines.

Morey's biographers say that he spent many months negotiating with Livingston to build steamboats for the Hudson, during which time Morey freely shared his plans and patents. When Livingston's final offer finally came, Morey was shocked. In return for signing over his patent rights, Morey would receive stock—not cash—with a potential value of $100,000. He would work as the assistant general manager of the company, and as if to sweeten this raw deal, the first boat would be named the *Lady Morey*. Miffed, he wanted no part of it and countered with an offer to sell Livingston his steamboat patents for $15,000 cash. Livingston offered half the amount, and then only if the scheme succeeded. Morey indignantly refused, and that was the end of that. When Morey saw Livingston and Fulton's side-wheeled steamboat years later on the Hudson, he reportedly said, "Blast his belly! He stole my patent!" But his point was moot; for whatever reason, Morey had never patented his paddlewheel design.[5]

In late 1797, Livingston and Stevens formed a partnership with Nicholas Roosevelt (whose brother James was Theodore Roosevelt's great-grandfather) to build a steamboat. Blood being thicker than water, Roosevelt's share in this deal was a stingy 12 percent of the patent rights, even though the patent was awarded in all three names; Livingston paid Roosevelt to build the boat and engine, however.[6] Livingston chose Roosevelt because he was one of the few men in America with steam engine experience. In 1793, Roosevelt had partnered with Philip Schuyler to rebuild Josiah Hornblower's 1750s Newcomen steam engine at Schuyler's copper mine at Belleville, New Jersey.[7] Roosevelt later took over an iron foundry and machine shop across the Passaic River from the mine and boldly named the operation Soho Works, after Boulton

& Watt's famous manufacturing plant in Birmingham, England. In 1798, Roosevelt hired two English engineers who had worked for Boulton & Watt, James Smallman and John Hewitt, to build Watt-type steam engines there, including the engine for the partners' new steamboat, *Polacca*.

Livingston, a lawyer by education, had supreme confidence in his engineering abilities. To make matters worse, he was, in today's parlance, an unrelenting micromanager. He and Roosevelt were soon squabbling about how to mount the paddlewheels. Charles Stoudinger, Roosevelt's young German mechanic, agreed with his boss's plan to use two vertical side paddlewheels.[8] But Livingston was set on his pet concept, a horizontal wheel mounted under the keel that forced water through an aperture in the stern.[9] Because he was paying the bills, he prevailed. Progress on the engine moved so slowly that year that at one point Livingston threatened to buy an engine from Boulton & Watt. Several months later he did write to the company, though nothing came of it.[10]

In an early trial of the *Polacca*, Livingston's horizontal wheel failed miserably. He continued to reject Roosevelt's suggestion for side paddlewheels, but he allowed Stevens to mount what he termed "elliptical paddles"—similar to Fitch's design of a decade earlier—at the stern. With those in place, in October 1798 they tried again. The boat pushed forward a short distance and then all hell broke loose. The pipes split at the seams and spewed hot water everywhere, and the boat's planking was ripped apart by the violent vibrations of the engine. Stevens determined that the problem was the Watt-type engine, designed for use on land. It shook so hard that the connection pipes broke. Livingston threw in the towel, at least for a while.[11]

Two years later, Livingston, Roosevelt, and Stevens agreed to build a stronger, redesigned steamboat. They signed a twenty-year contract in which each would carry one-third of the costs and receive one-third of the profits. Their second boat was tried sometime in late 1800 or early 1801, but its engine also caused excessive vibrations and was scrapped.[12] Not long after, President Jefferson appointed Livingston to serve as minister to France.

By the time Livingston arrived in Paris in late 1801, Robert Fulton, still searching for fame and fortune, had been living there since 1797. Now thirty-six years old, Fulton had moved there from England, where he had spent several years working, unsuccessfully, on canal design schemes. Before that, this farmer's son from Lancaster, Pennsylvania, had spent six years trying to make a career as a miniature-portrait artist. He had left Philadelphia for London in the fall of 1786 or spring of 1787, just months before Fitch tried his first steamboat on the Delaware.

His first few years in England as an artist were discouraging. Fulton finally got his first break in 1791 when he received a commission to paint a portrait of Viscount William Courtenay, a wealthy young nobleman and well-known homosexual in Devonshire. Fulton, at the time twenty-five, lived in Courtenay's castle for more than a year before moving to nearby Torquay. Around this time he began to realize that his career as an artist was going nowhere, and he began to look for other ways to make a gentleman's living. His observations of the work of some local craftsmen led to his first invention, a machine for cutting and polishing marble.[13]

He also became intrigued with local projects involving canals and came up with several ideas for new types of locks. Canals were beginning to earn good profits for their owners in England at the time, and Fulton saw a chance to turn ideas into money. In September 1793, he wrote to Charles Mahon, the third earl of Stanhope—the same Stanhope who visited Rumsey's steamboat in London—explaining some of his canal ideas. Stanhope at the time was serving as chairman of the Bude Canal company, which was in the planning stages. Rather audaciously for a person with no previous engineering experience, Fulton suggested to Stanhope that there were better ways to build canals. Stanhope responded by knocking down most of Fulton's canal ideas one by one, but the two wrote back and forth for another month. Their correspondence ended with a withering letter from Stanhope to Fulton: "I doubt whether you will do well to pursue mechaniks at present as a profession."[14] Fulton didn't let that criticism discourage him.

Fulton spent the next three and a half years studying canals and proposing schemes in northern England. In the process he invented a canal-digging machine and designed inclined plane systems, for which he received English patents. He continued to try to interest Stanhope in his ideas, but nothing ever came of his attempts. Exactly how he supported himself is unknown. At one point he was paid well for preparing a report for a Manchester canal company, which was published in 1796 with the title "A Treatise on the Improvement of Canal Navigation." The company officers must have felt hoodwinked when they finally read it, for it was largely an argument for how a system of networked small canals could benefit America.[15] (He sent a copy to George Washington, who replied with a polite thank-you note.)[16] Fulton's other ideas never brought him much money, but here and there he was able to sell interests in his schemes.

In the spring or summer of 1797 he left for Paris, ostensibly to obtain patents for his canal devices. He also hoped to determine if his system of small, networked canals would work in France. Then he planned to continue on home to America, where canal building was beginning to take off.

Fulton took a room at Madame Hillaire's pension on the Left Bank, a favorite haunt of Americans in Paris. He became friendly with one of the residents, a woman nearly ten years his senior named Ruth Baldwin Barlow. Ruth's husband, Joel Barlow—yes, the same Barlow who patented James Rumsey's tubular boiler in France—was serving as U.S. consul to Algiers. When Barlow returned to Paris in September, the three became fast friends and took a large apartment together. Fulton loved Paris. He wrote to a friend in England that "Paris is all gay and joyous as if there were no war at all."[17]

While waiting for his patents to be granted, Fulton tossed around several ideas for inventions, but the one that grabbed his imagination was a machine that could navigate underwater—a submarine, or what he called a "plunging machine." The idea was not original. The concept had been around for at least two hundred years, and a model had been tested in London in 1620. In 1776, David Bushnell built the *American Turtle,* an underwater vessel he used to try to plant a mine under a British ship, not entirely successfully, in New York harbor during the Revolution. Interestingly, Bushnell and Barlow attended Yale University at the same time, and they had several mutual friends, so Barlow could have given Fulton the idea. Some of the *Turtle*'s design elements were present in Fulton's *Nautilus,* but Fulton always claimed the invention was his alone, even after Bushnell's plans were published in the *Transactions* of the American Philosophical Society in 1799.[18]

Fulton's version was shaped like a fat cigar and measured about twenty-one feet long and six feet wide. An iron tank held water for ballast, and water was pumped in and out of it with a hand crank, allowing the crew to take on or release water and thus control their ascent and descent. The vessel moved forward when the crew turned a hand-cranked propeller. It could stay submerged for up to three hours.

In December 1797, his ideas barely developed, Fulton sent a proposal for a submarine to the French government. He described, without providing much detail, how his invention could be used to sneak up under enemy ships and plant torpedoes, another idea also probably borrowed from Bushnell. These were devices that we would call underwater mines, water-tight copper cylinders packed with gunpowder and equipped with timers or detonators to give the sailors in the submarine time to crank away to a safe distance. One of his conditions was that the French promise not to use his devices against American ships, a fairly naïve demand considering that the United States and France were close to war around this time, a situation Fulton must have been aware of.[19]

Intrigued, the French marine ministry studied Fulton's proposal for a month or so and then turned him down, partly because he was asking a lot of money for an unproven technology and partly because many French officers thought such a weapon was too horrible and immoral to use. Fulton didn't let this rejection discourage him. When a change in ministry leadership occurred six months later, he submitted another proposal, this time accompanied by rough plans and a scale model. A committee was named to look into the idea and not long after recommended that the government go forward with it. Fulton wasted no time in drawing up an agreement. In the spirit of defense contractors for decades to come, Fulton offered France this killer machine for the bargain price of 500,000 francs for every British vessel it blew up—more than $100,000 in the U.S. currency of the day. The marine ministry never responded.[20]

Fulton was furious at being ignored and wrote letter after letter to try to talk the French into a contract. After a few months of waiting for a reply, he gave up and began to explore other lucrative ideas. One was a rope-making machine along the lines of one developed by an English inventor and friend of his, Edmund Cartwright, a pioneer in power textile loom development. Since France allowed patents of importation, Fulton had no qualms about building on Cartwright's idea and obtaining a patent. He even found a U.S. partner to finance and build the machine.[21]

He then borrowed another idea, a design for a panorama, for which he received a French patent in April 1799. Ten years earlier, an artist in Edinburgh created a painting mounted on the walls of a round room, twenty-five feet in diameter that, when viewed from the center of the room, offered a 360-degree view of the Scottish city. It proved so popular that a similar room, called a panorama, was built in London in 1792; Fulton probably saw it while he was living in England. He bought a piece of land on Paris's Boulevard Montmartre and built a structure that contained a room forty-six feet in diameter. He hired painters to create a scene of Paris. The panorama was a huge hit, and Fulton was soon flush with the proceeds from admission charges. He sold the building later that year in a deal that gave him a share of the profits for the life of the patent, fifteen years. A second panorama was built next to the original, and the two attractions would make money for years.[22] (Today a brass plaque marks their location on the Passage des Panoramas, now a shopping arcade.)

At some point Fulton returned to his submarine and torpedo plans. After several more proposals and threats to take his invention to another country, in March 1801 he finally got the French government to sign a contract with him for a payment of 10,000 francs, far less than he originally demanded and a fraction of what he had already spent on the project. The future take would be great, however, with generous payments for each ship destroyed. By this time he had built a working submarine, which he had tested to an incredulous crowd one June day in 1800 on the Seine. With two men at the controls, the vessel stayed submerged in the river for twenty minutes at a time.[23]

In tests conducted that summer against British ships off the coast, the *Nautilus* never proved its worth. The little submarine was too slow and too clumsy to maneuver close enough to the enemy. Fulton gave up on the submarine and proceeded to deliver his torpedoes with regular boats and courageous divers. That didn't work either. By late summer he wrote to the government that things hadn't panned out as expected, but he had a better idea: build much larger submarines that could slip into British harbors and drop mines, then leave.

When Napoleon asked to see the *Nautilus,* Fulton told him he had destroyed it because it leaked, a lie to keep the French from copying his design. Then Fulton demanded that from then on that the French pay for all construction costs. These remarks did not sit well with Napoleon. A month or so later, the Peace of Amiens was signed, putting on hold the hostilities between England and France. Fulton shelved his submarine plans and settled in for the winter of 1801–1802 with the Barlows in the crumbling mansion Joel had bought for a song near the Luxembourg Gardens.[24]

In March 1802, Livingston and Fulton met. Livingston was just six months into his new job as minister to France, and at the time he was about to receive instructions from President Jefferson to begin talks with the French government about ceding or selling New Orleans and the Floridas to the United States.[25] Napoleon had secretly acquired the Louisiana Territory in 1800 from Spain, which had been eager to get rid of it for financial reasons and had happily traded it for Tuscany. Napoleon was hoping to establish a new empire in North America, using profits from the Caribbean sugar trade. This made Jefferson nervous, which led to his sending instructions to Livingston.

Even with this weighty subject on his mind, Livingston began talking with Fulton about steamboats. They soon were making plans to build a boat together in Paris, and both agreed that a ready-made English steam engine was key. The two men must have laughed when they discovered that each had written letters to Boulton & Watt in the past, and that neither had ever received a reply. (Boulton, we can only imagine, must have had a file folder labeled "Crazy Steamboat Projectors.") They decided to look elsewhere for an engine. In the meantime, Livingston hired the French mechanic Étienne Calla to build a model boat, powered by clockwork springs, that Fulton could use to test various propulsion methods.[26]

While Calla worked on the model, Fulton and Ruth Barlow left Paris in April for the resort spa at Plombières, in the Vosges mountains, where it was hoped a soak in the mineral waters would help heal a skin condition afflicting her genital area. Fulton's later biographers seem to agree that Fulton and Ruth were lovers, with Barlow's full approval. These same biographers are not quite so sure that the two men had some sort of sexual relationship as well, but it was clear that the three were quite fond of one another. Barlow wrote love letters to his wife in a silly baby-talk language, often referring to Fulton as "Toot." By this time, the threesome had been living together for nearly five years.

The boat model, three feet long and eight inches at the beam, arrived sometime in late May, and Fulton blocked a small stream on the grounds of the resort to create a pond for his experiments. Interestingly, Fulton found that the best method for moving the boat—better than oars, jet propulsion, or waterwheels—was an endless chain of paddles, ironically the first design John Fitch proposed but never built. Fulton wrote a paper summing up his experiments in which he made note of "Messrs. Parker & Rumsies experiment for moving boats." From this, it seems that either he saw Rumsey's boat firsthand while in London ten years earlier or had obtained a detailed description from Daniel Parker, Rumsey's partner and Barlow's friend. Fulton concluded that the jet-propulsion method failed because it "appears that the engine was not loaded to its full power, that the water was lifted four times too high and that the tube by which the water escaped was more than five times too small."[27]

While Fulton experimented, Barlow and Parker visited a Paris museum called the National Depot of Machines (today's Conservatoire National des Arts et Metiérs) and wrote to Fulton, "[T]here I saw a strange thing; it was no less than your very steamboat in all its parts and principles in a very elegant model. It contains your wheel-oars precisely as you have placed them, except that it has four wheels on each side to guide round the endless chain instead of two. The two upper wheels seem to be only to support the chain; perhaps it is an improvement."[28] This model was built by a man named Desblancs, a watchmaker who worked on the unsuccessful steamboat developed by Jouffroy d'Abbans on the River Saône near Lyon in 1783.[29] In April 1802, Desblancs obtained a patent and tried his boat on the Saône, but it failed as well.

Perhaps to avoid infringing on Desblanc's patent, Fulton switched to paddlewheels. When Livingston got word of this change, he told Barlow to tell

Fulton that the idea could not be patented in America because someone had beaten him to it; he must have been thinking of Morey or Roosevelt. But Fulton didn't let this news change his plans. He figured that patent law would protect his particular design, even if the wheels themselves were nothing new.

Fulton and Ruth spent the summer at the resort, returning to Paris in September. The following month he and Livingston signed an agreement that established a company to take out a U.S. patent for "a boat 120 feet long, 8 feet wide, and 15 inches draught . . . to carry sixty passengers, and to run between New York and Albany at the rate of eight miles per hour in still water."[30] The two men, after some hesitation on Fulton's part, agreed to share the costs and profits down the middle. Livingston didn't let the fact that he had a twenty-year agreement with Stevens and Roosevelt to build steamboats interfere with this new contract.

That winter, workmen were hired to build a full-size boat for trials to be conducted on the Seine, near today's Pont de l'Alma, not far from where the Eiffel Tower now stands. They ordered or borrowed a steam engine from the mechanic Jacques Périer; plans to buy an English engine never panned out because of the difficulty in getting an export permit. Meanwhile, Livingston managed to get his friends back in the New York Assembly to extend his monopoly for another two years, giving him until April 1807 to get a steamboat running on the Hudson.

Late one night in May 1803, the same month Livingston and James Monroe signed the Louisiana Purchase treaty, Fulton was roused from his sleep by friends who told him his boat had been smashed and sunk. He spent the rest of that night and much of the following day raising the engine and salvaging what he could. Reports vary on the cause, from a violent storm to a deliberate act of destruction. Part of the problem may have been that the engine was too heavy for the boat.[31]

Fulton started over, this time building a larger, stronger boat, seventy-five feet long and eight feet wide. At six o'clock in the evening of August 9, with invited dignitaries and a large crowd in attendance, the new boat's boiler was fired up. A Paris newspaper reported, "Fulton put his boat in motion with two other boats in tow behind it, and for an hour and a half he afforded the curious spectacle of a boat moved by wheels like a cart, these wheels being provided with paddles or flat plates and being moved by a fire engine." It went almost three miles an hour against the current—a disappointment, but Fulton was sure that with a Watt engine he could obtain the needed four-mile-an-hour speed required by the New York monopoly.[32]

In fact, three days before the public trial he wrote to Boulton & Watt instructing the company to build him an engine of twenty-four horsepower and prepare it for shipment to New York. He failed to mention that he planned to use it in a boat, no doubt aware of Watt's low opinion of steam-powered vessels. The letter was filled with questions: Would a different furnace be needed to burn wood instead of coal? What might happen to the condenser if brackish water was used? How large would the boiler have to be for this size engine?

He received a reply in September, and he quickly sent a letter back answering their questions and urging Boulton & Watt to proceed without further instructions, "as the communication between this country and England is daily growing more difficult." The two countries were again at war. A month later Fulton received bad news: Boulton & Watt declined his order for lack of an export permit. Fulton immediately wrote to James Monroe, who had just arrived as U.S. minister to Great Britain, and asked for his help. Fulton warned Monroe that it might be best not to use his name, since the British might recall his work for the French in trying to blow up their navy a few years earlier. Monroe never replied. He sent another letter to Monroe in November, with the same result.[33] The progress of the steamboat stalled once more.

━■━■━■━■━■━■━

Around the time of his first steamboat trial in Paris, Fulton was approached in confidence by an agent of the British government to build his submarines and torpedoes in England. This secret mission was the result of a request by none other than Fulton's former correspondent, the earl of Stanhope. Stanhope was serving in Parliament at the time and had learned of Fulton's submarine. In a closed session of the House of Lords in 1802, he warned the members that the device posed a threat to British shipping and that Fulton needed to be stopped and brought back to England.[34]

Fulton named his price, which was exorbitant, but after a few more rounds of talks a deal was set. Fulton crossed the channel in April 1804 using the alias Robert Francis, leaving his friends with the impression that the trip was merely a stopover on his way to America. When he arrived in London, he was told that the royal navy was interested not so much in his submarine as in his underwater mines. Fulton agreed to build an arsenal and test the devices on French ships, which were by then grouping off the coast, apparently preparing for an invasion of Britain. He demanded a salary of £200 a month, expenses of £7,000, and huge awards for each ship destroyed, for a period of fourteen years. While he was at it, Fulton, in a brilliant stroke, asked for and received permission from the British government to have a steam engine built for use in the United States; he later got permission to export it as well. Once he had the permit in hand, Fulton immediately traveled to Birmingham and met with Boulton personally.[35]

From the start, the British government failed to get its money's worth. That fall, protected by the darkness of night, sailors on rafts carried Fulton's mines to French vessels sitting in the harbors of Boulogne and Calais and planned to tie them to the anchor ropes. But the mines floated off and exploded onshore, hurting no one. The British finally put a stop to the project but kept Fulton on salary, a move that angered him even more. He demanded to be allowed to have another try, or he would demand a payment of £100,000 to quit the business altogether. They gave him one last chance on two French ships in September 1805. The mines exploded, but the ships were undamaged. In desperation, Fulton asked for the chance to blow up a captured ship. The government agreed, and he finally succeeded, but only because the ship was anchored and undefended and the act carried out in broad daylight—hardly warlike conditions. Nevertheless, the British paid him the agreed-upon £10,000 for this one-time test.

The following month, Horatio Nelson's forces soundly defeated the French at Trafalgar, and the threat of invasion was over. Fulton's services were no longer needed. The British government stopped his salary a few months later, but Fulton continued to demand more money, at one point asking for £60,000 up front and £200 a month for life in exchange for not selling his weapons system to another country. He promised that if he broke that agreement later, he would relinquish his salary but keep the lump payment.

In January 1806 he complained in a letter to William Pitt the Younger, the prime minister, that he had come to England "to show that I had the power and the might, in the exercise of my plan to acquire fortune, to do you an infinite injury." There was no response; Pitt was ill at the time and died later that month.[36] Fulton soon flooded the mailboxes of the new administration with more demanding letters and threats, hinting strongly that if a settlement wasn't reached he would be free to make the details of his discoveries public, as well as make public all correspondence on the matter that had passed between him and the government.

After months of Fulton's harassment, the British government finally felt it had to get rid of him. For a much smaller amount, about £1,600 over what he had already been paid, Fulton was sent on his way. He was furious. As he prepared to leave England in September 1806, he wrote to Barlow, who had returned with Ruth to America a year earlier, that he was free to "burn, sink, and destroy whom I please, and I shall now seriously set about giving liberty to the seas by publishing my system of attack."[37]

Before heading home, Fulton updated his wardrobe. Spending hundreds of pounds of his newfound wealth, he purchased twenty-three shirts, fifty cravats, twenty-seven pocket handkerchiefs, twenty-one waistcoats, eleven pairs of breeches, five coats, and three great coats, among other items.[38] After living in Europe for nearly twenty years, Fulton returned to the United States in December 1806. His custom-made Boulton & Watt steam engine had been sitting in a New York warehouse for more than a year.

While Fulton was busy finding ways to sink ships, several other men had pushed forward with steamboats and steam engines, both in Britain and the United States. In 1801, the engineer William Symington, who had built a disappointing steamboat in Scotland in 1788 for Patrick Miller, was back in business working for Lord Dundas, a major investor in the Clyde and Forth canal. His lordship hoped to use steamboat tugs instead of horses to pull canal barges. Symington mounted a paddlewheel at the stern and constructed a ten-horsepower, low-pressure steam engine that featured a piston rod attached to a crankshaft by a connecting rod, the first known use of this efficient method for obtaining rotary motion.

In its first test in March 1802, the *Charlotte Dundas,* named for his lordship's daughter, successfully pulled two seventy-ton barges for nearly twenty miles in six hours. Lord Dundas's canal partners, though, were concerned that the boat's paddlewheel would create waves that would cause the banks to wash away, so they refused to allow its use on the canal. Soon after, the duke of Bridgewater talked with Symington about building eight steamboat tugs for use on his canal, but he died before the agreement could be finalized. Symington never built another steamboat.[39] Reports that Fulton saw the *Charlotte Dundas* in Scotland are probably not true.

In the United States, John Stevens in Hoboken and Oliver Evans in Philadelphia were both experimenting with high-pressure steam engines, something Boulton & Watt firmly refused to manufacture because of the tendency of the boilers to explode. Sometime in 1802, John Stevens read about Evans's tests in Philadelphia and asked his brother-in-law, Dr. John Coxe, who lived there, to stop in Evans's workshops and ask a few questions. Evans had been aware of Stevens's attempts at steamboats and willingly shared his knowledge. Their devices seemed to be quite different, but to protect his claim of priority, Stevens sent a detailed description of his plans to a Philadelphia scientific journal.[40] Later, though, each man accused the other of stealing his ideas.

Evans had been working on steam engines for longer than anyone in America. He claimed that he first thought about using steam to move wheeled vehicles back in 1772, when he was a teenager apprenticed to a wagon maker. His first attempt to obtain patent rights for steam wagons, as he called them, was in 1786 in Pennsylvania, where the assembly thought him "insane" and refused his request. The following year the state of Maryland gave him a patent, figuring that there was no harm in doing so. Around that time he also began to talk to others about the use of steam engines to power boats with paddlewheels. But he could never find anyone "willing to contribute to the expense, or even to encourage me to risque it myself."[41]

Evans, born in Newport, Delaware, in 1755, is widely considered the nation's first truly great inventor. He became fairly well known in the 1780s as the inventor of the world's first completely automated mill. In his revolutionary design, the mill's waterwheel worked a series of interconnected belts, gears, elevators, rakes, and shafts, eliminating the need for human intervention in the process and producing finer, drier flour to boot. Excited by the labor-saving prospects this mill provided, Evans was sure millers everywhere would line up to buy his licenses for its design. He was wrong. Some distrusted the mill's complicated workings, while others, including Thomas Jefferson, saw the advantages but complained about having to pay him a licensing fee. Evans received one of the first U.S. patents in 1790 for his mill designs.

In 1802, he returned to his steam engine ideas. He built his—and America's—first high-pressure steam engine in his shops at the corner of Market and Ninth Streets, Philadelphia, where Gimbel's department store would later stand. He used it to grind plaster of Paris and saw marble, receiving a U.S. patent for its design in February 1804. After that success, he was soon selling high-pressure steam engines to owners of grist mills and saw mills all over the country. Besides being much more powerful than Watt-type steam engines, his were smaller, lighter, and easier to build.

Evans, excited by the prospects of such a powerful machine, again began to seek encouragement and financial support for his boat and land carriage ideas. One man he confided in was the architect and engineer Benjamin Henry Latrobe, a British immigrant and engineer who had recently completed construction of the Philadelphia waterworks. This plant supplied water to city residents by pumping water from the Delaware River using two huge low-pressure engines built by Nicholas Roosevelt at his Soho Works.[42]

Latrobe thought Evans's ideas for steamboats and steam wagons were a little far-fetched, and he obliquely referred to Evans in a paper he presented to the American Philosophical Society in May 1803, entitled "First Report in Answer to the Inquiry 'Whether any, and what improvements have been made in the construction of steam engines in America?'"[43] Latrobe complained in this paper, "A sort of mania began to prevail, which indeed has not entirely subsided, for impelling boats by steam-engines." He listed the following problems with the engines of the day: the weight of the engine and fuel; the large space occupied; the tendency of their action to rock the vessel and render it leaky; the expense of maintenance; the irregularity of their motion, and the motion of the water in the boiler and cistern, and of the fuel, with the vessel in rough water; and the difficulty arising from the liability of the paddles or oars to break, if light; and from the weight, if made too strong.[44] Evans agreed with Latrobe that these problems plagued any boat equipped with Watt-type low-pressure engines, but he resented Latrobe calling those who advocated steamboats and steam vehicles "crackbrained."

Two years later, Evans got his first chance to build a steam-powered vehicle when he received a city contract to build a river dredger. In his Centre Square workshops, on the spot where the famous Wanamaker's department store would be built later that century, he built a fifteen-ton flat-bottomed scow that was thirty feet long and twelve feet wide. A high-pressure steam engine powered the dredging machinery as well as a stern-mounted paddlewheel. He called this odd machine the *Orukter Amphibilos,* a Greek way of saying "amphibious digger."

To show the world that steam-powered land vehicles were possible, he attached four heavy-duty wheels to the scow. He somehow connected the wheels to the steam engine, and for several days in July 1805 drove it around Centre Square for all to see. One morning he rolled it out of his shop and propelled this lumbering machine past excited and fearful crowds the mile and a half to the Schuylkill River. He rolled it down an embankment into the water, cut loose the wheels, hooked up the paddlewheel to the steam engine, and steamed over to the Delaware River "leaving all the vessels going up behind me.[45] This vehicle was America's first "car" and the first working steamboat on the Delaware since John Fitch's *Perseverance* fifteen years earlier.

While Evans worked in Philadelphia, John Stevens had given up on his partners Roosevelt and Livingston and began building his own engines and steamboats in Hoboken. One of his first trials caught the attention of some students at Columbia College across the river in New York as they walked in Battery Park on a May day in 1804: "As we entered the gate from Broadway, we saw . . . a crowd running toward the river. On inquiring the cause, we were informed that "Jack" Stevens was going over to Hoboken in a queer sort of boat. On reaching the bulkhead by which the Battery was then bounded, we saw lying against it a vessel about the size of a Whitehall row-boat, in which there was a small engine *but no visible means of propulsion.*[46]

Stevens named this small boat *Little Juliana,* after one of his daughters. The means of propulsion was a first for a steam-powered boat: twin-screw propellers mounted at the stern. He wasn't the first to think about screw pro-pellers—Archimedes and Bernoulli knew their potential, and Bushnell and Fulton had used hand-cranked propellers on their submarines. Early on, Fitch tried hand-turned twin propellers on a small boat but couldn't get them to work well enough.

Two years earlier, Stevens had tried a single propeller, powered by a sim-ple steam engine of his own design—patented in 1803, it was direct acting and noncondensing—on a small flat-bottomed craft. He described the pro-peller as having "wings like those on the arms of a windmill." He was able to make a speed of four miles an hour. But the single propeller made the boat tend to go in circles, creating a steering problem. The *Little Juliana* was his next attempt, with an improved multi-tubular boiler and twin-screw pro-pellers, which turned in opposite directions to ensure straight sailing.[47]

Stevens knew he was on to something. He began to work on a larger boat that could be put into commercial service. Abiding by his contract with Liv-ingston and Roosevelt, he kept both men informed about his progress. But before he could get another steamboat running on the Hudson, Robert Ful-ton came along and stole the show.

TWELVE

STEAMBOAT COLLISIONS

The day will come when some more powerful man will get fame and riches from my invention.

—*John Fitch*

Looking like a sawmill mounted on a raft and set afire, as one eyewitness described it, the world's first commercially successful steamboat began its maiden crawl up the Hudson River on Monday, August 17, 1807. With little fanfare in the press—who could have known the significance?—the narrow boat, propelled by two huge side paddlewheels, moved upstream at the breakneck speed of four miles an hour. Heavy black smoke spewed from its stack. A sea monster, some observers called it disparagingly.[1]

Fitch and Rumsey would not have believed how quickly Robert Fulton put his steamboat into service. In less than six months in 1807, and only a few weeks after returning to the United States, Fulton hired a boatbuilder, retrieved his imported Boulton & Watt engine from a warehouse, and supervised an assortment of craftsmen and laborers to pull the whole thing together. Just a week after making a short trial run, he made his first trip from New York City to Albany and back without a hitch.

Fulton, of course, was operating under several advantages Fitch and Rumsey had lacked. He had plenty of his own money, the result of his extortionary deals with the British government. In Livingston, he had a business partner with even more money and political power to boot. He bought a ready-made, reliable steam engine, sidestepping most of the problems that dragged down the previous inventors. He was able to examine the work of several other steamboat builders, including Fitch and Rumsey. Thanks to Thornton, now head of the U.S. Patent Office, he was able to study the dozen or so U.S. steam-related patents that had come before. Then he set his methodical mind to figuring out what did and did not work. As Fulton freely admitted, he never really invented the steamboat. Rather, he built the first steamboat that really worked.

Fulton arrived in New York City in December 1806, pausing to drop a note to his partner Livingston—apparently the first time they had communicated in at least two years—to tell him he was ready to resume steamboat building.[2] The city had grown considerably during the two decades Fulton was in Europe and was now America's largest, with about ninety thousand people, nearly double the size of Philadelphia. Fulton didn't linger there, eager as he was to get to Philadelphia for a happy reunion with Ruth and Joel Barlow. The Barlows had returned to America the previous year, and Joel was busy attending to the printing of his epic poem on America, "Columbiad."

In January, Fulton and Barlow traveled to Washington, which had become a sociable town of about eight thousand in its sixth year as the nation's capital. Jefferson was in his second term as president, and Fulton and Barlow attended his New Year's reception at the White House. Later in the month the two were invited to the social event of the season, the banquet honoring Meriwether Lewis on the occasion of his return from his expedition west. Barlow introduced Fulton to the political and social scene, sought out old friends, and made a few new ones, in particular William Thornton.

Thornton had moved to Washington in 1794, not long after the collapse of Fitch's Steamboat Company in Philadelphia. After winning the design competition for the U.S. Capitol in 1793, Thornton was named by President Washington as one of three city commissioners. His job was to oversee the construction and funding of government buildings.[3]

After the federal government moved from Philadelphia to Washington City in 1800, President Jefferson disbanded the city commission and gave Thornton two new jobs: magistrate of the District of Columbia and commissioner of bankruptcy. In 1802, the Patent Office became a separate bureau of the State Department, and Jefferson named Thornton its first clerk—although Thornton soon took to calling himself its superintendent. It was a job he would hold for the rest of his life.[4] Architecture remained a side interest, however, and by the time he met Fulton he had designed several private homes in Washington, including the Octagon House and a private Georgetown residence known as Tudor Place, both of which still stand.

A week after Fulton and Barlow met Thornton, Thornton invited them to his home. Shortly after that, Thornton, impressed with the inventor, allowed Fulton to examine all the steamboat patents that had been awarded to that time. Fulton didn't think much of what he saw; he wrote to Livingston that "not one . . . approaches practicality." [5]

But Fulton had other things on his mind that month in Washington. He spent most of his time promoting his first love, the explosive devices he called torpedoes. His timing could not have been better. The government was fearful that the British navy might attack U.S. ports at any time, and several plans were being considered as to how best defend them. Fulton met on several occasions with Secretary of State Madison and Secretary of the Navy Robert Smith to persuade them that his mines would cost much less, and could be put to use much sooner, than new warships or elaborate harbor defenses. A few months later Fulton would get the go-ahead and funds to build and demonstrate his torpedoes. No one in the government seemed the least concerned that Fulton had been selling this same technology to the French and the British, nor did anyone appear to have moral qualms about what might well be called the first weapons of mass destruction.

While he waited for approval for his torpedo project, Fulton set off for New York to sort out steamboat plans with Livingston. Their March meeting at Livingston's elegant estate, Clermont, was more contentious than friendly. Livingston was peeved that Fulton had taken his time getting back to America and to boatbuilding, knowing that the terms of his last extension on his New York monopoly required a steamboat to be in operation the following month, something that obviously was not going to happen. Livingston was also put off by Fulton's eagerness to launch steamboats on the Mississippi River, when it was his beloved Hudson River that had been his focus all along.

Tensions rose when Livingston seemed to recall a change to the steamboat agreement they had signed in 1802 in Paris. He was sure that instead of sharing expenses and profits fifty-fifty, they had amended the contract to give Livingston one-fourth of the Hudson River steamboat "free of all expense."[6] Fulton was taken aback; he couldn't recall any such change. Livingston hunted around his office but couldn't find his copy of any later contract. Without proof, Livingston finally had to agree to the fifty-fifty deal, including expenses and profits, that the original contract specified—unless, of course, the amended contract turned up. They then drew up a new contract, similar to the first but with terms that reduced Livingston's liability if the boat failed. Now all Livingston had to do was start paying the bills, since Fulton had paid for everything so far. More important, Livingston had to finagle yet another monopoly extension from the New York legislature.

Fulton returned to New York City and began hiring contractors. The hull, which would be 146 feet long and 12 feet wide, with a flat bottom and straight sides, was constructed at Corlear's Hook, on the East River. He paid the duties on his Boulton & Watt steam engine and had it moved from the customs house. By the end of March, things were going well enough for Fulton to make a quick trip to Philadelphia to comfort Ruth, whose brother had died unexpectedly.

Around this time, apparently acting on a suggestion made by Joel Barlow, President Jefferson asked Fulton to design a canal that would connect the Mississippi River with Lake Pontchartrain, part of a plan to defend the port of New Orleans. Refusing to even discuss the matter, Fulton quickly turned Jefferson down, saying he was too busy.[7]

By the end of April 1807, Fulton was back in New York City managing his steamboat's construction. A frequent visitor to the shipyard, full of questions, was John Stevens, who was still trying to build steamboats in New Jersey. Stevens told Fulton that he planned to run a commercial steamboat back and forth to Albany, regardless of Livingston's monopoly. He didn't foresee any problems, he said, for after all his sister was Livingston's wife, and the steamboat contract that he, Livingston, and Roosevelt had signed in 1800 was in force for another thirteen years. Fulton shrugged off his remarks, saying that the Livingston-Stevens contract didn't affect him. He also probably figured that anyone who had been trying for nearly two decades to get a steamboat to work didn't pose much of a threat. In fact, he seemed to enjoy talking with Stevens and freely answered his questions.

Fulton's close supervision kept the boat's construction on schedule, and by May he was confident enough to set July 20 as the day of its first test run. He estimated that the Manhattan–Albany operation would net the partners about $32,000 a year, assuming fifty passengers a trip, four trips a week, forty weeks a year. This was an enormous sum of money for the time, an amount, as he wrote Livingston in late May, that "makes me feel warm and generous."[8]

In late June, an international incident disrupted Fulton's steamboat plans. The British ship H.M.S. *Leopard* attacked the U.S.S. *Chesapeake* in the open sea after leaving Norfolk, killing three Americans and wounding eighteen; several others were captured. War seemed imminent, and Fulton decided this would be an ideal time to demonstrate his destructive devices to navy officials. Scrapping the July 20 date for his first steamboat trial, he instead set that day for a torpedo demonstration in New York harbor.

He found an old brig and brought it to anchor between Ellis and Governors Islands. At the appointed hour of one o'clock in the afternoon, hundreds of people gathered to watch from the Manhattan shoreline. Fulton directed his boatmen to row out to the brig and drop two mines—cylinders packed with perhaps a hundred pounds of gunpowder—tied to each end of 120-foot rope, near the anchor line. They were attached to cork floats. The idea was that the current would sweep the bombs underneath the brig, where they would strike the hull on either side, setting off a detonator that would cause them to explode. As the oarsmen furiously paddled away from the ship, onlookers covered their ears. But the blast didn't come. Fulton rowed out to the brig and found that the gunpowder had spilled out of the cylinders. He refilled them and directed the men to try again. This time the mines exploded well before they reached the boat, sending plumes of water into the air. Fulton brought out two more mines and started over. Progress was slow. By seven in the evening, most of the spectators had given up and gone home. But eventually the mines hit their target and the brig was ripped to shreds.

The New York press had a field day with Fulton's difficulties. Washington Irving, in his periodical *Salmagundi,* satirically reported that Fulton's underwater mines were "an excellent plan of defense; no need of batteries, forts, frigates, and gun-boats; observe, sir, all that's necessary is that the ships must come to anchor in a convenient place; watch must be asleep . . . fair wind and tide—*no moonlight*—machines well-directed—mustn't flash in the pan—bang's the word, and the vessel's blown up in a moment."[9]

Unfazed by near failure, Fulton wrote to President Jefferson that the demonstration was a success, and that the problems encountered had been noted and fixed. Jefferson, who was not a fan of a large navy, replied, "I consider your torpedoes as very valuable means of defense of harbors, and have no doubt that we should adopt them to a considerable degree." He added that he hoped Fulton had not given up on the submarine as a better means for delivering the mines. He closed the letter by once more urging Fulton to go to New Orleans to oversee the canal planned there.[10]

Fulton again ignored the president's call to duty and returned to taking care of the final details for his steamboat trial. On August 7, four years to the day after his successful steamboat run in Paris, he fired up the boiler and took the boat out on the East River for about a mile. The test showed that the force of the engine was so great that he could double the size of the paddle blades. He wrote to Livingston about this success and set August 17 as the date of his first run to Albany, a distance of about 150 miles.

With only Fulton, the captain, and the crew on board, the steamboat left New York about one in the afternoon on the seventeenth, paddled upriver for more than a hundred miles, and reached Clermont, Livingston's home, exactly twenty-four hours later. Fulton spent the night there. Although the boat was rigged with sails, Fulton didn't use them. He figured his speed averaged about four and a half miles per hour. The following morning around nine, Fulton resumed the journey, with Livingston on board, for the forty-mile trip to Albany. They arrived without incident, and without ceremony, around five in the afternoon. They had made the trip in thirty-two hours of river time, a trip that normally took four days by sail.

Exhausted, Fulton spent the next two nights in Albany. On Wednesday, he hung a sign on the vessel announcing the boat's return to Manhattan the next morning, ready for its first paying passengers. Although the townsfolk showed much interest in the strange vessel, only two brave souls, visiting Frenchmen, made the trip. This time the word of the steamboat's journey spread downriver, and people lined the shores to watch its return to New York. But the New York newspapers, their pages filled that summer with reports of Aaron Burr's conspiracy trial, had little to say, other than to print a brief report Fulton sent them citing times and distances. Fulton told Barlow about the event:

> My steamboat voyage to Albany and back has turned out rather more favor-ably than I had calculated. . . . I overtook many sloops and schooners, beat-ing to the windward, and parted them as if they had been at anchor. The power of propelling boats by steam has now been fully proved. The morning I left New York, there were not perhaps thirty persons in the city who believed that the boat would ever move one mile per hour, or be of the least utility, and while we were putting off from the wharf, which was crowded with spectators, I heard a number of sarcastic remarks. This is the way ignorant men compli-ment what they call philosophers and projectors. . . . It will give a cheap and quick conveyance to the merchandise on the Mississippi, Missouri, and other great rivers, which are now laying open their treasures to the enterprises of our countrymen; and although the prospect of personal emolument has been some inducement to me, yet I feel infinitely more pleasure in reflecting on the immense advantage that my country will derive from the invention.[11]

Fulton spent the next two weeks getting his steamboat ready for regular passenger service, adding sleeping berths to the cabins, boarding the sides, and building covers for the boiler and engine works. He began dreaming of the fortunes to be made on the Mississippi. Forgetting his partner's resistance to the idea, he asked Livingston in late August to get detailed information from his brother, Edward, who lived in New Orleans, about river velocity, traffic, cargo tonnage, and more.

On September 4, Fulton began regular passenger service of the *North River Steam Boat;* in those days, the lower part of the Hudson River was referred to as the North River. Passengers boarded at the Paulus (also spelled Powles) Hook Ferry dock, at the foot of Cortlandt Street. Service continued on a regular basis for another two months, until ice began to form upriver. Service was interrupted only a few times for minor repairs, most of which were caused by collisions—some apparently intentional—with sloops. On one occasion, a paddlewheel was damaged and had to be removed to make repairs. Such incidents would increase in the years ahead, initiated by boat owners who felt that their livelihoods were being threatened by Fulton's steamboats. By 1811, a state law had to be enacted to deal with these malicious acts.[12]

By the end of the *North River*'s first short season, Fulton reported to Livingston that they had cleared a 5 percent profit and estimated future profits of "8 to 10,000 dollars a year."[13] The steamboat's failure, so far, to explode, sink, or otherwise endanger passengers encouraged more people to give it a try, and soon as many as ninety people were making the trip each run. Fulton ordered some major changes to be made over the winter to the vessel, including strengthening the hull, widening the boat to sixteen feet to improve stability, adding new cabins and a new boiler, and installing luxurious touches, such as mahogany woodwork. Livingston, in turn, proposed cutting costs and suggested that the new boiler be made of wood. Fulton insisted on copper.

In early November, Fulton was back in Washington. Barlow had finally found a suitable home for the threesome, a thirty-acre estate above Rock Creek that he named Kalorama, Greek for "beautiful view." (The house, which stood near today's Twenty-third and S Streets, N.W., is gone, but the neighborhood northwest of Dupont Circle retains the name.) Fulton and the Barlows were happy to be settling in together at last, and they began making plans to remodel the house, at one point calling in Benjamin Latrobe for advice.

Around this time, Fulton began to think seriously about applying for a U.S. patent. Because the courts had never ruled if federal patent law invalidated existing state patents and state monopolies, Livingston, the lawyer, was concerned that applying for a patent might open a can of worms. Fulton agreed but felt that their only threat on the Hudson River would be from a patented boat of completely different design, and he didn't foresee that happening for a long time. Fulton decided that it would be best to obtain a patent for his boat design before anyone else did, to protect him from imitators in all the states.

Fulton must have frowned a few times as he read through a copy of the Patent Act of 1793 he had borrowed from Thornton. Section 1 began by stating that patents would be granted for "any new or useful art, machine . . . or any new and useful improvement . . . not known or used before the application." Fulton may have smiled at the word "improvement" until he read Section 2 and learned that "any person who shall have discovered an improvement in the principle of any machine . . . which shall have been patented . . . shall not be at liberty to make, use, or vend the original discovery." It was followed by the line that dashed his last hope: "simply changing the form or the proportions of any machine . . . in any degree, shall not be deemed a discovery."[14] The only good news, if there was any, was that the 1793 law had discarded the stringent patent examination process. Now, any applicant who met all the administrative requirements, swore an oath of originality, and paid the fee would receive a patent. Any infringements or controversies would have to be resolved in court. A patent could be repealed if it was determined it was not legally made.

Fulton faced a challenge: how to write his application to make a convincing argument for the originality of his invention. He asked Livingston for help: "[A]id me with your ideas of the distinction which makes the novelty of this invention."[15] Livingston offered nothing. Frustrated, Fulton set the problem aside for a while. Remaining in Washington with the Barlows, he continued to push the government to fund further demonstrations of his torpedoes. He wrote a detailed letter to Jefferson in early December laying out costs and plans.[16]

Livingston, in the meantime, was busy lobbying his friends in the New York legislature to extend his steamboat monopoly. By the following spring, the extension was approved, and Fulton and Livingston gained five additional years for each of the next two boats they put into service, for a maximum of ten more years on top of their current monopoly. This was plenty of time for both men to make fortunes.

Around Christmas, Fulton prepared to return to New York to oversee the improvements being made to the *North River*. Just before leaving Washington, he received a strange letter from Thornton announcing that he was planning to take out patents for steamboat inventions. He noted that Fitch's 1791 steamboat patent, which he claimed contained several of his ideas, had expired two years earlier. He offered to share these patents with Fulton and Livingston, with each party paying one-third of the costs and receiving one-third of the profits of any boat they built together.

Thornton told Fulton that furthermore he possessed, as a director and largest stockholder in Fitch's defunct Steamboat Company, a document signed by the king of Spain giving the company rights to the navigation of the Mississippi River. This claim was more than a slight exaggeration. It seems Thornton was referring to a permit issued to Fitch's company by the Spanish governor of New Orleans in 1791 allowing their steamboat to go upriver, albeit without cargo.[17] No record of any other document pertaining to Mississippi River rights for Fitch's company has ever been referred to or found. In any case, it is hard to imagine that rights issued by a country that no longer controlled the Mississippi would be honored years later by the U.S. government. Fulton must have chuckled at Thornton's impudence; he didn't think the letter worthy of a reply.

After a brief layover in New York City, Fulton headed north on January 5, 1808, for Teviotdale, another Livingston family estate not far from Clermont. Two days later, seemingly out of the blue, he married Harriet Livingston, a cousin of the chancellor's. It is unclear when the two met—Livingston family legend claims the relationship began when Fulton docked overnight at Clermont on his steamboat's maiden voyage. If a courtship followed, no mention of it has ever been found.

Harriet, who lived with her widowed mother and other family members at Teviotdale, was twenty-four years old (Fulton was then forty-two), reasonably attractive, well-educated, and well-set financially. Fulton perhaps wanted to project a conventional image now that he was back in conservative America. Other than Ruth, he never seemed to show much interest in women, and in fact had talked of marriage only once before. That occurred shortly before he left England, when he wrote to Barlow that he was thinking of marrying a wealthy widow there. Barlow shot back an emotional letter pleading with Fulton to reconsider, predicting that the woman would never want to live in America and that he and Ruth would never see him again. Fulton returned alone.[18]

As it turned out, January 7, the day of Fulton's wedding, was also memorable for John Stevens, who now was nearly sixty years old. That day he signed a contract with a Hoboken shipbuilder for a full-size steamboat, to be named the *Phoenix,* measuring one hundred feet long and sixteen feet wide. His experiments the previous fall with a fifteen-foot-long steamboat, with side-mounted paddlewheels replacing his earlier screw propellers, had encouraged him to proceed with a full-size boat. The craft reached a speed of six miles per hour, using a Watt-type steam engine that Stevens had adapted to use high pressure.[19]

Fulton and Livingston were concerned but not surprised when they got the news of Stevens's plans. They had been talking with Stevens for several months about forming a partnership because they didn't want to fight him on the Hudson. Stevens, on the other hand, was angry that Livingston continued to ignore their 1800 steamboat contract. He also believed that Livingston's New York monopoly was unfair and possibly unconstitutional given the federal patent law. Fulton and Livingston made Stevens several offers. One gave Stevens permission to copy Fulton's plan on the condition that Stevens acknowledged that he was working under Fulton's patent.[20] Stevens, who possessed, along with Fitch and Rumsey, one of the first steamboat patents, was too proud to admit the superiority of the upstart Fulton. Another offer gave Stevens permission to run steam ferries and steamboats on specified, noncompeting routes—but certainly not the route up the Hudson River to Albany. Stevens had long planned to set his boats in service on that river. He either ignored or refused every offer the two men presented. But he kept his options open and remained friendly with both.

A week or so before he signed the contract to build the *Phoenix,* Stevens made a partnership move of his own. He asked Livingston in a letter if they could try once more to work together, saying it was his "sincere wish to unite upon equitable principles."[21] Livingston and Fulton replied in mid-January 1808 with a twenty-page letter in which they rehashed their previous conditions and tediously defended the Hudson River steamboat monopoly. Livingston argued that state monopoly rights and federal patent law were two entirely different things. Fulton blasted Stevens for blatantly copying the *North River,* from its proportions to its use of side paddlewheels; he argued that should Stevens build such a boat, Fulton's federal patent would protect him. It was quite a letter, legalistic in language and haughty in tone. Stevens's biographer summarized it as follows: "Take your boats out of your frontyard, the Hudson River, so that we can fill it with our boats. Establish yourself on the Delaware, at an inconvenient distance from Hoboken where you have located your building-ways and your shops. Admit that we are inventors and that you are no such thing. These conditions being met, we will protect you with our United States patent."[22] Stevens politely declined, noting he thought state monopolies would eventually be ruled unconstitutional and that he preferred to go it alone for the time being. Later Stevens was surprised to learn that despite what Fulton said about having a patent, he still had not submitted an application.

By the summer of 1808, Fulton and Livingston began sending threatening letters to Stevens, promising to have the *Phoenix* seized, as their New York monopoly allowed, as well as to bring suit against him for patent infringement. Stevens ignored them. Livingston then tried a more direct approach, licensing his brother, John R. Livingston, to run steamboats and steam ferries on the very New York and New Jersey routes he had previously offered to Stevens.[23] Around that time, Fulton and Livingston were dismayed to hear about more competition on the horizon. A Canadian steamboat was being built for service on Lake Champlain, and a group of investors in Albany was forming a steamboat company that planned to run on the Hudson River in defiance of the monopoly.

In late September 1808, after several construction delays, Stevens took the *Phoenix* on her first run, from Paulus Hook to Perth Amboy, a distance of about thirty miles. He estimated her speed to be about five and a half miles per hour, better than that of the *North River* and just as he had predicted. But the journey proved too strenuous for the machinery, and the boat had to be taken in for repairs. It would not run for the rest of the year.

Three months later, Fulton, now quite worried about the threat Stevens posed, challenged him publicly in an odd sort of way. He placed a letter in a New York newspaper haughtily asking Stevens to prove that he understood hydraulics, giving him a problem to solve and have printed in the same paper within two days' time. "If the answer does not then appear," Fulton taunted, "it is a fair inference that you are still in a state of experiment and working without any certain rule." Stevens refused, saying he would rather work with real boats than play games with pen and paper.[24]

The refurbished *North River* was ready for service in late April. It had no trouble attracting passengers. Because the boat had been substantially rebuilt, Fulton was required to re-register it. This time he expanded the name, calling it the *North River Steam Boat of Clermont.* For reasons unknown, in the years that followed the boat was almost always referred to as the *Clermont,* even though Fulton never called it by that name. By July, profits were running about $1,000 a week.[25] The many elegant touches Fulton added, including three separate sleeping cabins that could accommodate more than fifty passengers and a kitchen serving all meals, led to a long list of written rules: no feet on the furniture, no boots on the beds, and no smoking except in designated areas. There were charges for damage to the furniture and fines for rules violations, with the proceeds spent on wine for the other passengers. Other than complaints about excessive fines and inconvenient schedules, the operation began to run fairly smoothly. In late July, Fulton turned the day-to-day operations over to his employees and made plans to take Harriet to Washington, so they could live together at last with his friends at Kalorama.

One wonders how much Harriet really knew about the relationship between her husband and the Barlows. Her conservative upbringing in rural New York would not have meshed easily with the liberal attitudes of Paris, where Fulton, Joel, and Ruth spent so many happy years together. Before leaving New York, Fulton wrote to Ruth Barlow and told her how excited he was to be returning. "Shall we unite our fortunes to make Kalorama the center of taste, beauty, love, and dearest friendship . . . ?"

Predictably, the new living arrangement was an immediate disaster, especially for the women. Ruth, surely jealous of the new Mrs. Fulton, took to her room a good deal of the time. Harriet, five months pregnant, was uncomfortable around Ruth and unhappy to be so far from her family. Several weeks into the arrangement, Barlow escaped for a visit with Jefferson at Monticello, leaving Fulton to deal with the women. In October, Harriet gave birth to their first child, whom Fulton named Robert Barlow and called Barlow. That same month, Fulton contracted with the *North River*'s shipbuilder to begin the *Car of Neptune,* a larger and more luxurious version of the *North River.* With those two achievements behind him, he spent the remainder of 1808 at Kalorama struggling with his patent application. By the end of the year, the "household had turned in on itself," and the presence of the Fulton infant had brought all social activities at the mansion to a halt.[26]

On January 19, 1809, according to a letter of Thornton's (the only known mention of a date; all Patent Office records and models were destroyed in the Patent Office fire of 1836), Fulton delivered his patent application to Thornton's office. It had been professionally copied from Fulton's draft by a man named Fletcher and neatly packaged in a custom-made mahogany box.[27] As befitting the artist in him, Fulton's application was a triumph of embellishment, full of charts, diagrams, and sketches. Underneath all the glitz, the weakness of his claim to novelty could not be disguised. In the end, all he could come up with was the relationship between paddle-wheel size, hull size, bow angle, and a few other factors, which he claimed were the secret to his success. In other words, his application relied on proportions, a factor specifically cited in the Patent Act of 1793 as not patentable. There was also a technical problem with the patent: Fulton's signature was not in Fulton's hand but in Fletcher's.[28] Thornton, having corresponded frequently with Fulton, surely noticed and made a mental note for future reference.

Thornton had been expecting Fulton to deliver his application for several months. Just three days before Fulton showed up, Thornton, as threatened, issued himself a patent for steamboats and boilers. Thornton explained several years later that he had been waiting for Fitch's 1791 patent to expire (which it did in August 1805) so that he could take out his own patent for the ideas he had contributed to Fitch's boat and engine. He never explained why it took him more than three years to get around to it.[29]

The previous November, Thornton, failing to hear from Fulton on his partnership offer, began courting Stevens. Thornton's attempts to work with Fulton or Stevens had less to do with ego and more to do with money. Over the years, his debts had increased greatly, in some cases for reasons not entirely his fault, and his income declined dramatically. His family's plantation in Tortola was no longer a reliable source of cash, and his Patent Office salary was just $1,400 a year (he had received $2,000 a year as a city commissioner), hardly enough to support the socially active Washington lifestyle he and Anna Maria enjoyed. His expenses had increased as well, especially after he bought a farm in what is now downtown Bethesda, Maryland, and began breeding racehorses.

He wrote to Stevens and mentioned that Fulton was working on his patent application, a topic that grabbed Stevens's attention.[30] Stevens wrote back in early January asking Thornton to send him a copy as soon as Fulton filed it.[31] Later that month, Thornton replied, telling Stevens that he had taken out his own patents, that Fulton had brought in his application on January 19, and that he would ask Fulton for permission to make a copy.[32] This news alarmed Stevens, who quickly wrote back saying he thought patents were public records and that anyone willing to pay the costs could ask for a copy. In the meantime, Thornton told Fulton of Stevens's request, and Fulton adamantly refused to give permission.[33] Thornton relayed this news to a letter to Stevens in mid-February, explaining that the prohibition on making copies without the patentee's permission was a Patent Office custom and not an official State Department rule. Thornton later asked his boss, Secretary of State Madison, for guidance; Madison in turn asked the attorney general.

Fulton was unaware of Thornton's overtures to Stevens that winter. He was again busy promoting his torpedo projects and in January invited President Jefferson, President-elect Madison, and several members of Congress to Kalorama for a demonstration of his latest idea, a harpoon gun for planting torpedoes on a ship's hull. When their support failed to materialize, he returned to what some would call his treasonous ways. In a letter to a friend in the French government, he offered to rid the English Channel of British ships for two million francs.[34] America may have been close to war with Great Britain, but France was hardly a friend; French ships were still looting and seizing American vessels at sea.

With his patent application delivered and his weapons plans on hold, Fulton, Harriet, and little Barlow—surely at Harriet's urging—moved out of Kalorama in late February 1809. They returned to New York City, where Fulton rented a large house at 75 Chambers Street, near City Hall.[35] Joel Barlow was unhappy with Fulton for leaving, but Ruth did not protest the move.

In May, Thornton stirred the pot some more, writing to Fulton that Stevens had again requested a copy of his patent, and that he would continue to refuse that request unless the attorney general ruled otherwise. Then, in an attempt to revive the partnership issue, he launched into a lengthy critique of Fulton's patent, which he had issued on February 11. He told Fulton that paddlewheels on boats were certainly nothing new, having been around since the time of Isaac Newton. He reminded him that the patent law prohibited the patenting of existing designs, from any place or any time. He also warned against Fulton's reliance on proportions, noting that patent law expressly excluded proportions as patentable improvements. "How then would you defend your claims?" Thornton asked. "You cannot place any reliance on its defensibility." Thornton went on, telling Fulton that even if proportions were patentable, they simply copied Fitch's long, narrow boat of 1790: "Our boat was exactly eight feet wide and sixty feet keel—yours twenty by 150, each equal in length to 7–1/2 times the breadth!"[36]

Thornton sent a second letter a month later, repeating many of his criticisms and arguments, and concluding, "I have expended several thousand dollars on steamboats and am convinced I can build them with success." Fulton responded on May 19, disagreeing with Thornton and claiming he was the first and only person to take a scientific approach to developing steamboats. He ignored Thornton's request to form a partnership.[37]

Meanwhile, in New Jersey, Stevens finally had all he could take of Livingston's threats and Fulton's harassment. The *Phoenix* had begun its first full season of ferry service when Stevens heard that the *Raritan,* the steam ferry John R. Livingston operated under license from his brother and Fulton, would match the *Phoenix*'s schedule between New Brunswick and New York City—day for day and hour for hour, at a fare that was one-third less than that of the *Phoenix.* Stevens gave up the fight and made plans to move his operation to Philadelphia.[38]

To do so meant taking his river steamboat into the Atlantic Ocean for a voyage of about 150 miles, a huge risk, even supplemented by its two sails. Most observers considered the idea sheer lunacy. Stevens's second oldest son, the ironically named Robert Livingston Stevens, was eager to take the helm. Only twenty-one years old and already showing the signs of mechanical genius that would later make him famous, Robert Stevens had helped his father build the *Phoenix.* After days of waiting for good weather, Robert steamed out of New York on June 10, 1809, and headed for sea. But the weather quickly turned bad, and for much of the trip the *Phoenix* was forced to drop its anchor and ride out thunderstorms, lightning, high waves, swells, and heavy fog. Only three of its thirteen days at sea were fair. During the occasional calms, the paddlewheels kept her moving. By the time the boat approached Philadelphia, the boiler had suffered so much damage that Robert had to let her drift toward the Market Street wharf. Regardless, the *Phoenix* claimed its place in history as the world's first ocean-going steamer.[39]

Around the time the *Phoenix* left New York, Thornton wrote to an old friend in Philadelphia, Henry Voigt, who was still chief coiner for the U.S. Mint. He told him about Fulton's patent and asked Voigt's opinion. Voigt answered in late June that he thought Fulton's claim of being the first to use paddlewheels was debatable, saying, "I well remember paddle-wheels, I think the first use was in England." Voigt told Thornton that John Stevens or one of his representatives had approached him "some time ago" with similar questions about Fulton's boat. But to Voigt the bigger news was the recent arrival of the *Phoenix.* He told Thornton that the boat made its first round-trip between Philadelphia and Trenton on July 10, taking eight hours each way.[40]

That fall, Fulton and Livingston tried once more to reach a settlement with Stevens. "When relatives and friends grow cold toward each other, the cause is always to be lamented, yet when such an unfortunate event occurs, there is more nobleness of soul in nourishing a spirit of reconciliation, than in persevering in error," the partners began. They said they had heard his high-pressure engine was not performing well, nor was his tubular boiler, and that his stated plans to redesign his works would directly infringe on their patent. But, they wrote, they were ready to compromise.

This time, instead of asking Stevens to acknowledge publicly that he was working under their patent, they asked that he make such a statement privately. They saved their strongest argument for last: "Some gentlemen of ample funds have made application to us for permission to build two boats, on the Delaware, one to run between Philadelphia and Trenton; the other between Philadelphia and New Castle." If Stevens did not accept their latest offer, they said they would proceed to deal with this other party.[41]

Stevens's reply was predictable. "[N]ow I find that unless I am ready, *in forma pauperis,* to subscribe to terms the most humiliating and degrading, and against which my feelings revolt, I am probably doomed to be pursued and persecuted in like manner on the Delaware. . . . It is my firm conviction that it will ultimately prove more advantageous to me to stand the brunt of a legal contest with you, than to accommodate, even on my own terms." But he hesitated to burn his bridge completely. He closed with, "notwithstanding all that has passed, I am still cordially disposed to conciliate and accommodate on reasonable and equitable terms."[42]

The three eventually met in New York in December, and the outcome was an agreement signed by all three men giving Stevens the right to use the Fulton and Livingston patent on "the Delaware, Chesapeake, Santee, Savanna, and Connecticut Rivers, and from Rhode Island to Providence" so long as Stevens put at least one steamboat in operation on those waters within seven years. The agreement shut Stevens out of New York state and would not let him run steamboats between New York City and New Jersey or on the Ohio or Mississippi Rivers. It also gave Fulton and Livingston the right to use any of Stevens's improvements or inventions on their steamboats.[43]

With the Stevens problem at last taken care of, Fulton returned to working on weapons of war. He spent much of the winter of 1809–10 writing a sixty-page pamphlet entitled "Torpedo War and Submarine Explosions" and distributing it to influential people in Washington. (The title page bore his optimistic motto, "The Liberty of the Seas Will be the Happiness of the Earth.") The U.S. Navy commodore read it and was not impressed; he called Fulton's pamphlet the most impractical scheme "to have originated in the brain of a man not actually out of his mind."[44]

Throughout 1810 Fulton prepared for another demonstration of his torpedoes to the U.S. government; war with Britain seemed imminent. Congress had authorized Fulton $5,000 for further experiments, and in October of that year he carried out another spectacularly unsuccessful demonstration in New York. A congressional review of his efforts a few months later reported to him that the United States was no longer interested in his work. Falling back on his old ways, Fulton then wrote to the governments of France, Russia, and Holland offering to sell his weapons. There is no record of replies.

Turning back to steamboats, Fulton was increasingly worried that his patent would not protect him against future competition, so he began approaching states for monopoly rights. When Thornton heard about Fulton's attempts in Virginia and Ohio, he wrote letters to those state legislatures explaining that federal law would be rendered useless if state monopolies were granted to Fulton or anyone else. But the following year, the Fulton-Livingston team succeeded in winning the biggest prize of all: exclusive rights to the lower Mississippi River, granted by the governor of the Orleans Territory in April 1811. Again, Livingston's political connections handed Fulton an unbeatable advantage. By then, Fulton was close to running a steamboat on the Mississippi.

More than two years earlier, Benjamin Henry Latrobe, the engineer who had criticized steam engines on boats in 1803, had approached Fulton with a plea to give his new son-in-law, Nicholas Roosevelt, who was then out of work, some part of his steamboat business. In early February of 1809, Latrobe sat down at his desk in his Washington office to write a letter to Fulton about steamboat building in general and Roosevelt's qualifications in particular. The two men knew each other socially through their mutual friend, Joel Barlow. "I had hardly closed [the letter], when he, as if sent to me, walked into my office. I gave it to him."

As they talked things over that day, Latrobe reminded Fulton of Roosevelt's 1800 steamboat-building contract with Livingston and Stevens, which was still in force. But Fulton brushed it aside, saying that it did not affect him. Latrobe then mentioned Roosevelt's claim that he had urged Livingston, unsuccessfully, to use paddlewheels on their steamboat in 1798. Roosevelt was planning to take out a patent for his ideas, Latrobe went on, and he could prove he had priority for vertical paddlewheels. This was an attempt to poke a hole in Fulton's patent and use it as a polite threat to encourage them to work together. Fulton answered that there was no room for Roosevelt in the New York operation at the time, but that he was working on plans to organize a company for the Mississippi, and "into that I can let my friends."[45]

Latrobe was a professional architect and engineer (he had studied under John Smeaton, considered the father of British civil engineering). He had emigrated to America from England in 1796 shortly after his wife died. Landing in Norfolk, Virginia, he took on whatever architectural work he could find. His design work on two houses in Norfolk and Richmond and on the state penitentiary in Richmond caught the eye of the amateur architect Thomas Jefferson, whom he met in Fredericksburg.

After two years in Virginia, Latrobe, a tall, attractive, dark-haired man of thirty-four, moved to Philadelphia, where he had a better chance to work on lucrative projects. As the only formally trained architect in America at the time, he quickly landed the design award for the Bank of Pennsylvania, and the following year he won the contract for first Philadelphia waterworks, serving as both engineer and architect.

Philadelphia had suffered greatly from outbreaks of yellow fever every summer for several years, and Latrobe guessed that the city's contaminated drinking water was the source of the disease. He proposed two graceful buildings to house two steam-driven water pumps, one on the banks of the Schuylkill River and the other at Centre Square, the site of today's city hall. The idea was to pump water from the river to the middle of town, then distribute the water through underground pipes, in those days made of hollowed-out white pine logs. He hired Nicholas Roosevelt's Soho Works in New Jersey to build the huge steam engines for the pumps.

Latrobe and Roosevelt became close friends, and soon "Uncle Nick" was a frequent guest in the Latrobe household, which in 1801 consisted of Latrobe's new wife and his children, Lydia, age ten, and Henry, nine. Latrobe and Roosevelt both suffered tough times financially over the next few years, when payments for contracts, such as the waterworks, did not come through. Roosevelt had been awarded a navy contract to provide copper sheathing for a new fleet of frigates during the Adams administration. When Jefferson, who opposed expansion of the Navy, became president, he canceled the contract, leaving Roosevelt in financial straits, along with Latrobe, who had signed notes for the project.

In 1805, still struggling along, Roosevelt wrote a startling postscript in one of his business letters to Latrobe. He confessed that he was in love with Latrobe's daughter, Lydia, then thirteen, and that he wished to marry her. Latrobe went wild. At thirty-six, Roosevelt was old enough to be Lydia's father. Eventually, after much hand-wringing and worry on Latrobe's part, he gave in and allowed the two to marry, which they did after Lydia turned seventeen in 1808.[46]

Around that same time, Latrobe was fighting for his professional reputation, thanks to the cantankerous William Thornton. Latrobe first met Thornton not long after he came to America, when Thornton gave him a tour of the Capitol, then under construction. Now, nearly five years later, Latrobe, as second architect of the Capitol, was faced with correcting critical problems in Thornton's design. He made several attempts to meet with him, but Thornton refused to show up. Latrobe sent a report to Congress that noted the "absurdities of William Thornton's designs" and asked for funds to make needed changes.[47] Thornton took offense and began writing nasty letters about the architect to members of Congress and a Washington newspaper. Thornton's verbal attacks continued for several years and may even have become physical, as Latrobe's biographer suggests.[48] In 1808, Latrobe, fed up with Thornton's unceasing accusations and insults, sued him for libel. Interestingly, Thornton's lawyer, whose delaying tactics kept the lawsuit in limbo for several years, was Francis Scott Key.[49]

About a year after Fulton and Latrobe first talked about it, Fulton was ready to think about steamboats on the Mississippi. In June 1810, Fulton hired Roosevelt to make a survey of the 1,800-mile stretch of the Ohio and Mississippi Rivers, from Pittsburgh to New Orleans. If the trip revealed the rivers were navigable by a steamboat all the way, Fulton and Livingston would take on Roosevelt as their one-third partner in building a steamboat for the Mississippi—so long as he agreed to raise the money for it. Roosevelt set out that summer on a keelboat with measuring gear, a small crew, and his pregnant but adventurous wife. They spent the rest of the year floating down the two rivers. Lydia gave birth to their first child shortly after they returned to New York from New Orleans.

The Roosevelts' next trip down the Mississippi would not be so uneventful.

THIRTEEN

JOHN FITCH'S GHOST

It's the same each time with progress. First they ignore you, then they say you're mad, then dangerous, then there's a pause and then you can't find anyone who disagrees with you.

— *Tony Benn, British Labour politician*

Historians refer to 1811, the year a steamboat first wound its way down the Ohio and Mississippi Rivers, as the *annus mirabilis*— the wondrous year—of the West. Several unusual natural events, in addition to that one remarkable human one, had people talking about bad omens. In April, the Great Comet of 1811, one of the largest and brightest ever recorded, began its nine-month streak. That summer, the Mississippi and other rivers inexplicably overflowed their banks. Thousands of gray squirrels swarmed through the forests and across the Ohio River en masse, seemingly running (and swimming) for their lives. In September, a solar eclipse darkened the skies. In December, the weather in the region turned oppressively warm and still. Endless days of hazy skies made the sun look like a "glowing ball of copper."[1]

To cap off this strange year, on December 16, one of the largest earthquakes ever to hit the contiguous United States—to this day—devastated the central Mississippi valley. Fortunately the area was sparsely populated. The epicenter was north of the 400-person settlement of New Madrid, Missouri, on the banks of the Mississippi. Plaster fell off walls and chandeliers shook as far away as Virginia and South Carolina.[2]

Two days earlier, the captain of Roosevelt's steamboat, the *New Orleans,* had tied it up to the foot of a small island on the Ohio River above Yellow Bank (now Owensboro), Kentucky. The crew would spend the following day loading up firewood. Also on board the sky blue, 148-foot side-paddlewheeler were the pilot, a crew of six, Roosevelt and Lydia, a few maids, and a Newfoundland dog named Tiger. They had left Pittsburgh on October 20 and arrived in Louisville several days later. They waited there for a month for the water to rise high enough for the steamboat to pass over the Falls of the Ohio. Luckily, Lydia, who had been eight months' pregnant when they left Pittsburgh, gave birth to Henry, their second child, while they were stalled in Louisville.[3]

Shortly after two in the morning of December 16, the steamboat shuddered hard, followed by strange loud noise. Everyone on board woke up. It felt, they said later, as if the boat had run aground. But a quick check showed they were still in deep water. Those who could went back to sleep. When daylight came, they puzzled at the sight of a muddy, roiling river and hundreds of twisted and toppled trees on shore. The smell of sulfur filled the air. A little after eight o'clock, a second shock hit with an explosive roar, bigger than the first, and then they knew what had happened. The steamboat, partially shielded from turbulence and debris by the island behind them, pulled tight against its moorings but held. After several horrifying minutes, the shaking lessened. Roosevelt decided they would be safer on the water than on an unreliable shore, so he directed the crew to cast off and continue on. Once under steam, the noise and vibrations of the engine masked the continuing trembling far beneath them.

Roosevelt prayed they were heading away from the earthquake, but as they pressed on he realized otherwise. Huge chunks of riverbank had split and fallen into the river, along with the trees that had lined them. Empty flatboats and pieces of furniture floated by. Arriving later that day in the little settlement of Henderson, they were greeted by a frightened crowd. Roosevelt went ashore to see the damage. Most of the sturdy small log cabins were still standing, but all had lost their chimneys.

They pushed off again the next morning. The steamboat slid down the rolling river, somehow continuing to dodge trees and debris on its way toward Natchez. All those aboard, including the dog, remained anxious and mostly silent. For days, the aftershocks continued—one local man recorded eighty-nine strong tremors between December 17 and 23. Lydia wrote that she "lived in a constant fright, unable to sleep or sew, or read."

One night, after a day of relative calm, the captain again tied the steamboat to the foot of an island. Lydia had relaxed enough to be enjoying a rare sound sleep when the noise of objects striking and scraping the boat woke her up. The next morning, the pilot looked out from the deck, confused. At first nothing looked familiar. Had their cable snapped and sent them afloat? Then he spotted a few landmarks on the shore he remembered from the day before. He went to check the cable, and the boat was still tied to the same tree as the night before—except now the tree was underwater. The reality finally hit them. The steamboat hadn't moved, but the island had broken up. What Lydia heard that night was pieces of mud, rock, and vegetation bumping past.

Later that day they saw the most horrifying destruction of all. "At New Madrid . . . the earth opened in vast chasms and swallowed up houses and their inhabitants, terror-stricken people had begged to be taken on board, while others dreading the steamboat even more than the earthquake, hid themselves as she approached." The Roosevelts guiltily and regretfully turned down their pleas, explaining that they had barely enough food for themselves.

By the time the boat reached Natchez, five days before Christmas, the river began to look normal. At this town, a prosperous place that was the first capital of the Mississippi Territory, the *New Orleans* received a hero's welcome. Eleven days later, on January 10, 1812, the Roosevelts arrived in New Orleans to a surprised and cheering crowd. Few had expected the steamboat to complete its voyage, earthquakes notwithstanding.

Back in New York, Fulton was fuming. Word of the boat's success had reached the East, and the reports irritated him to no end. As Latrobe explained later,

"The newspapers were filled with accounts of "Mr. *Roosevelt's* boat," and Mr. *Roosevelt's* successful navigation, and Mr. *Roosevelt's* dinners on board, without a word said of Messieurs Livingston and Fulton."[4] Not only that, Fulton also complained he did not receive a single letter from Roosevelt during the entire journey, even though he had mandated regular reports. The cost overruns while the boat was being built in Pittsburgh, Roosevelt's sloppy accounting, his failure to raise enough subscriptions, and questions about unrelated business investments—made with Fulton's money?—all turned Fulton against his partner in the Mississippi Steamboat Navigation Company.

So Fulton took action. He refused to pay Roosevelt bills that had come due. He and Livingston, by a two-thirds vote, cut Roosevelt out of his share of future profits (which turned out to be substantial). He threatened Roosevelt with a lawsuit to recover company funds he suspected had been spent on other things. Finally, he refused to talk to Roosevelt, by letter or in person.

Latrobe stepped in to help, visiting Fulton in New York in late January 1812 to try to smooth things over. Latrobe told him his son-in-law could not have been responsible for the newspaper stories about the *New Orleans* because "he was on the river when they were published." He suggested that Roosevelt's reports to Fulton had been lost because of the earthquake. And Latrobe was sure, he argued, that Roosevelt was an honest man, just not very good at recordkeeping. Fulton, who respected Latrobe, relented. He eventually agreed to pay Roosevelt's bills, to not bring suit against him, and to restore the one-third share of the boat's profits—but only to Lydia and her heirs. That was all, Fulton said; he refused to deal with Roosevelt again. He directed John Livingston, his wife's brother, to take over the steamboat's planned service between Natchez and New Orleans and send Roosevelt packing.[5]

The year 1811 had been a tough one for Fulton. In January, fearful that his Hudson River monopoly would not restrain competitors, he submitted a second patent application. He hoped this one would be stronger, filled as it was with improvements he had made in two years of actual steamboat building. One of his self-described "innovations" was the discovery that fifty-ton boats were more efficient users of energy than smaller boats. Thornton, when he read it, guffawed—how could Fulton be so brazen as to try to keep others from building boats of a certain size?

Thornton, however, was required to issue patents so long as fees were paid and procedures followed. Before he issued this one, though, he wrote to Fulton and threatened to file an official protest against the application (in which case a board would decide on its validity) because it contained nothing new. He also teased Fulton with the fact that Fitch's 1790 steamboat had gone faster than any of Fulton's to date.

Fulton wrote back from Kalorama in January saying he was too ill to come see Thornton to discuss the matter but that his own experience convinced him that no steamboat using a Boulton & Watt type engine could ever exceed six miles an hour. He returned Thornton's taunt with a challenge: "To prove your principles by practice it has occurred to me that one of two things may be done, either that you find someone to join you with funds to build the boat and if you succeed to run 6 miles an hour in still water with one hundred tons of merchandise I will contract to reimburse the cost of the boat and give you 150 thousand dollars for your patent, or if you can convince me of the success by drawings or demonstrations I will join with you in the expenses and profits, please to think of this and have the goodness to let me see or hear from you as soon as possible."[6] Thornton failed to respond to this offer and issued Fulton his second patent on February 9, 1811.

By this time Thornton had given up on partnering with Fulton anyway. At the moment he was offering to share his patent with the Albany Company—the "22 pirates," as Fulton called the group of investors who dared to flout his monopoly on the Hudson.[7] But they didn't need Thornton's patent because they were blatantly copying Fulton's steamboat, and in fact construction had already begun on their two boats, the *Hope* and the *Perseverance*. In what could only be another swat at Fulton, Thornton, knowing that some sort of legal action would ensue, supported the Albany Company by sending its shareholders information they could use to argue the weaknesses of the Livingston-Fulton monopoly and Fulton's U.S. patents.[8]

The *Hope,* looking like a twin of the *North River,* began regular runs between New York City and Albany in June, going head to head with Fulton's newest steamboat, the *Car of Neptune,* as well as the *North River* and the

Paragon. The Albany Company's *Perseverance* was launched in September. On July 27, the first-ever steamboat race took place on the Hudson River between the *Hope* and the *North River.* Leaving Albany at the same time, the captains of the two boats pushed their engines hard to make a speed of five miles an hour. After a few miles, as one steamboat tried to pass the other, they collided. Noting with relief that there was little damage, the captains agreed to call the race a draw.

Fulton was indeed planning to force the Albany Company off the Hudson, and that summer he began gathering material for a legal fight. He wrote to his former acquaintance in England, the earl of Stanhope, asking him to certify that they had communicated on the topic of steamboats back in 1793 and that Fulton had specifically proposed side paddlewheels. For reasons unknown, Stanhope never replied, although a review of his papers years later showed he had received Fulton's request.[9]

Fulton also asked his friend Joel Barlow for help, writing to him at Kalorama in June. "A more infamous and outrageous attack upon mental property has not disgraced America. Thornton has been one of the great causes of it."[10] Fulton asked Barlow to get Thornton to sign a prepared deposition stating that Fulton was the first to use side paddlewheels and the first to devise several specific improvements and features. He also wanted Thornton to state that no steamboats had ever been or were then permanently operated in Europe.

Barlow was happy to carry out what would be his last favor for his old friend, even though he and Ruth were busy packing. He had just been named minister to France by President Madison, with the specific goal of getting Napoleon to provide restitution for the American ships France had seized over the years.[11] Fulton's instructions regarding Thornton were clear: "Get him to call on you, get him in a private room, no evidence. Point out to him in firm language the mischief which the Albanians [Fulton's term for the members of the Albany Company] vow and for which we will pursue him if he does not immediately do us all the justice in his power."

Barlow did his best but failed. He reported back to Fulton, "The poor fellow can depose nothing now unless it be his bones. He has not recovered from his fever and it is thought by some that he never will. . . . I called and took him out that morning in my carriage before breakfast and kept him at the judge's till eleven o'clock when I sent him home. It seems he was sick with fever when I took him out, tho' I did not know it. I leave your papers for him with Cutting who promises to make him attend to it as soon as possible."[12]

In July 1811, Livingston and Fulton went to a circuit court in New York to obtain an injunction against the Albany Company. The judge, Henry B. Livingston (yes, a cousin of the chancellor's), dismissed the case. Livingston and Fulton proceeded to state court, where they asked for triple damages and the right to seize the Albany Company's two boats. There, lawyers for the Albany Company argued, armed with goodies from Thornton, that Fulton's federal patent was invalid and that the New York state monopoly was unconstitutional because it restricted commerce. The judge agreed and refused the injunction. Fulton and Livingston then appealed to the court of errors, made up of the governor, two senators, and several judges—certainly a setup politically favorable to Livingston. Not surprisingly, when the court convened in March 1812, it ruled for Livingston and Fulton, noting that states had every right to confer monopolies.[13]

Rather than seize the two boats, which would not help his image among New Yorkers, Fulton made several attempts to negotiate a settlement with the Albany Company. When those failed, Fulton bought the *Hope,* and the *Perseverance* was sold to a group of investors that included a businessman and politician named Aaron Ogden. The fate of the *Perseverance* is unclear, but Ogden would begin his year as governor of New Jersey that November.[14]

■ ■ ■ ■ ■ ■ ■ ■ ■ ■

Latrobe was back in Fulton's New York office in the fall of 1812, this time to discuss a role for himself building steamboats. Times were tough, and the best and brightest architect and engineer in America was in desperate need of a job to support his growing family. President Madison had declared war on Great Britain in June, and Latrobe's work on the Capitol and other government buildings had ceased for lack of funds. To make matters worse, the economy was in shambles after months of British blockades of U.S. ports, which didn't help Latrobe's private architectural business.

Latrobe was a perfect choice for Fulton because of his previous experience with steam engines. After serving as contractor for the Philadelphia waterworks, Latrobe ordered an engine to be built for the Navy Yard in Washington. At the time, he and his son, Henry, had won the contract to build the New Orleans waterworks. When problems arose in dealing with his engine-maker, the former Boulton & Watt engineer James Smallman, Latrobe decided to establish his own engine-making shop in Washington. He planned to move the works to Pittsburgh, where he could put the engines on barges and float them downriver to New Orleans.[15] Getting Fulton to pay for the move would be a help.

Fulton's steamboat business was booming. When Latrobe showed up, Fulton was putting together the Ohio Steamboat Navigation Company and looking for someone he could trust to build the boats. In July 1813, the two men met again, and Latrobe signed on as a partner.[16]

Fulton knew he needed to move quickly to get steamboats on the Ohio. The hefty profits being made on the Hudson and the Mississippi had investors salivating to support steamboat companies, which were forming at a rapid pace. Thornton's public statements about the weakness of Fulton's patents further encouraged competition. A businessman named Daniel French already had a steamboat running on the Delaware and was preparing boats for the Ohio and Mississippi. Oliver Evans had announced plans to build steamboats powered by high-pressure engines on the Ohio. Both men were using or planning to use side paddlewheels in spite of Fulton's patent.[17]

Fulton and Evans had once carried on a friendly correspondence, but for some reason that ended after Fulton visited Evans at his Mars Works in Philadelphia in early 1811. Evans had been trying for a while to sell Fulton on the benefits of high-pressure steam engines. But Fulton was concerned about their safety and told him that low-pressure engines were good enough for his boats. In a nasty letter written after that visit, Fulton bet Evans that boats powered by his high-pressure engines would never go more than ten miles per hour. Evans wrote back and told Fulton that his patents were worthless.[18]

Before Latrobe moved to Pittsburgh in the fall of 1813, where he would build a steamboat at the yards of Fulton's Mississippi Navigation Company, he sold subscriptions in Washington for a steamboat Fulton was planning for the Potomac River.[19] Around that time Fulton hired a man named John Delacy, one of his attorneys in the fight against the Albany Company, to drum up subscribers for a steamboat to run between Richmond and Norfolk. Thinking globally, he also began inquiring about gaining rights to run steamboats in Russia and India.

In the middle of these expansion plans, Fulton lost his partner. In late February 1813, he received word that Livingston, who had been in failing health for several months, had died of a stroke at Clermont. He was sixty-six. Later the same day a letter arrived that knocked Fulton flat. Joel Barlow had died of pneumonia two months earlier in Poland, where he had been trying to track down Napoleon. For a time Fulton was so grief-stricken that he talked about giving up the steamboat business.[20]

Livingston's estate was a tangled mess, and Fulton spent months haggling with the various Livingston brothers and cousins over how to run the various steamboat enterprises. While they bickered, Aaron Ogden, whose year-long term as New Jersey governor had ended, exerted his political influence in the Livingston style and convinced his state legislature to give him a steamboat monopoly. That in hand, he made plans to run on the Hudson in direct competition with Fulton's and John R. Livingston's New York–New Jersey steam ferries.[21]

Aaron Ogden wasn't the only interloper on the Hudson. In defiance of his agreement with Livingston and Fulton, John Stevens had been running his newest steamboat, the *Juliana,* as a ferry between Hoboken and New York City for almost two years. Fulton ignored him for a while, but in the summer of 1813 he threatened Stevens with a lawsuit. Stevens moved the boat to Connecticut to keep the boat safe from seizure.[22] Fulton had been angry to learn that Stevens, also in violation of their agreement, had lobbied the legislatures of Virginia and North Carolina for steamboat monopolies. Virginia turned him down, but North Carolina had awarded him exclusive rights several months earlier. Fulton exploded when he learned that Stevens had used his Fulton-Livingston connection to obtain the monopoly. Fulton threatened Stevens with a lawsuit and sent Delacy to North Carolina to get the grant repealed, claiming it had been obtained "surreptitiously and by false suggestions."[23] Later that year, North Carolina revoked the monopoly.

Stevens, in the meantime, carried on. He launched another steamboat, the *Philadelphia,* on the Delaware, and he began building a horseboat to be run as a ferry out of Hoboken. John Fitch, who had begged him to consider this very idea years before, would not have been amused.

Latrobe moved his family from Washington to Pittsburgh in the fall of 1813 and found the bustling industrial town not much to his liking. He wrote to a friend a few months later, "Mud and smoke are the great evils of the town. Whoever can make up his mind to breathe dirt, and eat dirt, and be up to his knees in dirt, may live very happily and comfortably here."[24]

Fulton had estimated it would cost no more than $25,000 for Latrobe to build the Ohio River steamboat named *Buffalo.* But Latrobe incurred one large unexpected expense as soon as he got there. To his dismay, he found that the shops and wharves of Fulton's Mississippi Company, which he had planned to use, were at full capacity building John Livingston's steamboats, the *Etna* and the *Vesuvius.* Latrobe had to buy land and build and equip shops of his own, an added expense he said he cleared with Fulton.[25] As work progressed, parts and materials turned out to be more expensive than estimated, in many cases higher than in New York.

For the next several months and well into the following year, the story was the same. In March 1814, as costs for the *Buffalo* were nearing $25,000, Fulton sent Latrobe a letter saying he would refuse to pay any more notes written against him. After much pleading and explaining by Latrobe, Fulton gave in and sent more money. In July, the boat still wasn't finished, and Latrobe wrote to Fulton warning that the actual cost would now approach $45,000, counting the value of the shops and tools. Latrobe tried not to worry; John Livingston told him that his two steamboats, both slightly larger than the *Buffalo,* were coming in at about $45,000 each. Latrobe also had assumed from the start that the estimate for the *Buffalo* had not been meant as a fixed price.

Fulton's response to another request for more money was cold. "Fulton became angry[,] reproached me with benefits conferred on me and my family and left my arguments unnoticed," Latrobe said. "He accused me of speculating, and I stated all my private transactions [Latrobe took on architectural and construction work and was building steam engines for other customers while in Pittsburgh, but he claimed he put all his profits into the steamboat.]. . . . At last he turned me over to the company, and our correspondence ceased."[26]

With no more cash coming from Fulton, Latrobe desperately tried to raise the $7,000 needed to finish the boat. He appealed to the company's local shareholders on September 1. They refused to help. When he reported his problems to the Ohio Company directors in New York, they shot back a curt letter telling him to turn the steamboat and other property over to the Pittsburgh agents of the Mississippi Company. Latrobe, looking for a way to protect his project, talked to an attorney and learned that Pennsylvania law gave builders the right to place a lien on their works, which he then did. He then informed the Ohio company that he was prepared to settle things in a friendly way "unless treated in an [sic] hostile manner."[27]

Two weeks later Fulton fired Latrobe as his agent. A bitter war of letters ensued that lasted into November. No sooner did that battle end than another began. The Ohio Company's directors told Latrobe that Fulton had given *them* a fixed price for building the boats and that they planned to hold Fulton to that price. In the end, Latrobe kept his shops but lost the *Buffalo*.[28] He was out of work again.

■—■—■—■—■—■—■—■—■

Latrobe was only one of Fulton's many problems by 1814. Armed with his New Jersey monopoly, Aaron Ogden had been running his own steamboat ferry, the *Seahorse,* from Elizabethtown across the Hudson to New York City. In late January, he turned around and asked the New York legislature to repeal the Fulton-Livingston monopoly. Fulton hired two top lawyers, Thomas Emmet and Cadwallader Colden, to represent him at the committee hearings in Albany the following month. With ammunition from Thornton, Ogden laid out his case. In March, the committee found in favor of Ogden, saying that Fulton's steamboats "were in substance the invention of John Fitch."[29]

Fulton, who had been ill with lung problems that winter, had not been present at the hearings. The committee's ruling shocked him into action. He rushed to Albany and pleaded his case before the full assembly. He stressed that because of his and Livingston's investments and hard work, New Yorkers were privileged to have a quick and reliable form of transportation. Furthermore, it would be several years before he could make enough money on the Hudson to repay the debts incurred in building the four boats he had in service.

To prove his priority of invention, Fulton presented a copy of the letter he said he had sent to the earl of Stanhope in 1793 in which he proposed the use of side paddlewheels—five years before Roosevelt had the idea. After this oration, Emmet took the floor and made a long-winded and dramatic plea for Fulton, portraying him as an unappreciated and wounded hero. It worked. The assembly voted fifty-one to forty-three to deny Ogden's request, and the Fulton monopoly on the Hudson was saved.[30]

The War of 1812 shifted into high gear during the summer of 1814. In August, the British burned every public building in Washington, including the White House and the Capitol. The one exception was Blodgett's Hotel, which housed the Patent Office. Thornton hurried over in time to see British troops marching toward it. He approached the officer in charge and urged him to spare the building, telling him it contained hundreds of models and papers and arguing that "to burn what would be useful to all mankind, would be as barbarous as formerly to burn the Alexandrian Library, for which the Turks have been ever since condemned by all enlightened nations."[31] The colonel in charge considered that thought for a minute, then ordered the building spared. It later served as the temporary Capitol. (During that time of rebuilding, one of Thornton's first architectural commissions, the Octagon House, became the temporary White House.)

Fulton took time away from steamboats that summer to return to his true love, the defense business. At one point, Fulton was so taken up in military preparations that he asked President Madison to name him secretary of war. Setting the standard for defense contractors for years to come, he offered the federal government, for an upfront fee of $40,000 plus costs, the services of the *New Orleans,* the *Buffalo,* and his two almost-ready Mississippi steamboats, *Vesuvius* and *Etna,* to transport troops and supplies from Louisville to New Orleans, where British ships were massing.[32]

Baltimore was the next British target, and Fulton thought his torpedoes might be useful in defending the harbor. He made his way to Washington and headed straight to Kalorama, which Ruth Barlow was renting to the French ambassador. He retrieved several torpedoes he had stored there several years earlier and took them to Baltimore. The navy was ready to try anything, but after several attempts against British ships in Baltimore harbor, the torpedoes again failed to do the job. The "rocket's red glare" at Fort McHenry that September came from British Congreve rockets, not Fulton's torpedoes.

Fulton's war machines by now had progressed beyond torpedoes and submarines. Early in 1814 he signed a government contract to build one or more heavily fortified and armed steam frigates, for use in guarding city ports. In October, *Fulton I,* destined for the port of New York and built like a catamaran with a huge paddlewheel mounted in between two hulls, was launched from its East River shipyard and towed to Jersey City to have its machinery installed. Fulton took out a patent on the design but patriotically did not charge the government royalties, only the $240,000 it cost to build, plus a 10 percent commission, standard practice for the time.[33] The war ended early the following year, and the frigate, unfinished, never saw military action.

Fulton's Mississippi steamboats never carried troops or supplies, either; the *New Orleans* hit a snag and sank in July 1814, the *Buffalo* and the *Etna* were not ready in time, and the *Vesuvius* became stuck on a sandbar that summer. It remained there for six months, by which time the Battle of New Orleans was over. Fortunately for Fulton, he received his $40,000 government fee in late December. It appears he never returned the money.[34]

While all of this was going on, Fulton was facing another legal battle, again involving Aaron Ogden. Ogden was using his New Jersey monopoly to great effect against his competitor in the New York–New Jersey ferry business, John R. Livingston, who had been running a profitable ferry service under license from Fulton for several years. In a neat turnabout, John R. asked the New Jersey legislature to repeal Ogden's monopoly. This time, in addition to Thornton, Ogden could count on Nicholas Roosevelt and

John Delacy, two disgruntled former Fulton agents looking for revenge, to help out.

Despite Fulton's protest of his patent application, Roosevelt had received a U.S. patent on December 1 for his vertical side-paddlewheels and several other steamboat-related inventions (interestingly, three weeks later Thornton issued himself a patent for paddlewheels on steamboats).[35] Latrobe was ready to sign an affidavit stating that he was with Roosevelt when he had proposed paddlewheels to Livingston in 1798. Even though patent issues were unrelated to the monopoly question, Ogden was looking for ways to discredit Fulton. The Roosevelt paddlewheel priority issue was as good as any and in fact turned out to be a focus of the hearings.

This news sent Fulton to riffling through Livingston's correspondence for any evidence that his dead partner and Roosevelt had ever discussed the use of side paddlewheels, as Roosevelt had long claimed. To his dismay, he found several. He tucked them away and prayed that Roosevelt had failed to keep copies.[36] Five letters in particular, written in September and October 1798, contained references to side paddlewheels and vindicated Roosevelt. The first, written to Livingston from Passaic on September 6, was a progress report on the steamboat they were building together. Citing problems of insufficient power with Livingston's underwater horizontal wheels, Roosevelt said, "I would therefore recommend that we throw two wheels of wood over the side." Roosevelt kept pressing for his paddlewheels when finally Livingston put the matter to rest in his October 28 letter to Roosevelt: "As for vertical wheels they are out of the question."[37]

Fulton also assumed, correctly, that Thornton would somehow have a hand in the upcoming hearings, so he wrote a letter on December 27—not his first—to Secretary of State James Monroe complaining about Thornton's practice of issuing himself patents and using his office to secure deals for himself.[38] As it turned out, Monroe, who had been receiving similar complaints from other inventors, had that very day sent a letter to Thornton saying that effective February 15, 1815, the Patent Office superintendent could no longer issue himself patents or have a personal interest in any patents issued.[39] Thornton responded angrily, claiming the ruling represented "a great injustice on me" and that his interest in patents was necessary to supplement his meager Patent Office salary, which had not increased since 1802.[40] On the same January day that Thornton wrote his letter, the impatient Fulton sent a query to Richard Rush, the U.S. attorney general, asking what action the government had taken to restrict Thornton.[41]

In the meantime, Thornton revised and expanded a pamphlet he had written in 1810. He had it printed in Washington just before the New Jersey hearings were to begin. Titled "Short Account of the Origin of Steamboats," the twenty-page document discounted Fulton as the inventor of the steamboat and honored the achievements of Fitch. In it Thornton told the story of the Fitch-Rumsey fight, Fitch's attempt to build steamboats in France with Aaron Vail's help, Thornton's own involvement in Fitch's company, and his later dealings with Fulton. An appendix contained a copy of Fitch's patent application and a deposition from Oliver Evans stating that he had once "suggested to the said Fitch, the plan of driving or propelling the said boat by paddle or flutter wheel at the sides of the boat; then the said Fitch . . . informed him that one of the company had already proposed the use of wheels at the sides, but that he had rejected them." A final appendix item contained a bombshell that would be read at the hearings to great effect.

On January 14, 1815, the New Jersey legislature heard arguments on the case. Fulton was there, represented again by Thomas Emmet. Ogden hired Joseph Hopkinson and Samuel Southard as his attorneys. Delacy, who was representing Roosevelt, wrote to his client about the first day's hearings. He mentioned that the legality of Fulton's 1809 patent had come up. "Fulton has the effrontery to avow his having got Fletcher to sign his name and make light of it, as if he was entitled to violate the laws, as well as private rights, at pleasure."[42] The committee adjourned for a week to examine the evidence.

On January 23, the New Jersey state house was packed with curious spectators eager to witness a bitter fight among the rich and famous. Emmet, in an attempt to prove his client was the first true inventor of the steamboat, stated that John Fitch had played around with the idea but never got a steamboat up and running; he was probably trying to refute the earlier conclusion of the New York legislative committee that Fulton's steamboat was essentially the same as Fitch's. Ogden's lawyer, Southard, countered by taking the stand and reading from Fitch's journal and correspondence, citing evidence that Fitch had run several steamboats, including the 1790 boat that saw commercial service on the Delaware. He portrayed Fitch as a forgotten and unappreciated genius who died alone and in poverty. Thornton, to whom Fitch had turned over many of his letters and papers before leaving for France in 1793, was probably the source of these documents.[43]

Near the end of the following day, Ogden's lawyers called on Ferdinando Fairfax, heir to the vast Lord Fairfax estate in Virginia and Thornton's new partner in steamboat ventures, to testify. Fairfax read from a letter that Nathaniel Cutting sent him. Cutting, an American businessman, had partnered with Fulton in a French patent for a rope-making machine in Paris in the late 1790s. The business arrangement failed, partly due to Fulton's diversion of Cutting's investment. Fairfax had tracked down Cutting and offered him a chance to get even.[44]

Cutting said that while he was in Paris in 1805, he had a long conversation with Aaron Vail, who was still the American consul at Lorient, and that they began discussing Fulton's inventions. "Mr. Vail allowed that Mr. Fulton was not the first among our countrymen to have succeeded in propelling a boat by steam. . . . That Mr. Fitch came to France . . . but could not obtain the pecuniary aid requisite for his purpose . . . and deposited his specifications and drawings in the hands of Mr. Vail and quit the pursuit in France. . . . [Vail] lent to Mr. Fulton at Paris all the specifications and drawings of Mr. Fitch, and they remained in his possession for several months."[45]

Cutting concluded that Fulton had "adopted some of Mr. Fitch's improvements and hints without giving credit for them. My suspicion on this end was founded on a positive conviction of his having done the same thing with respect to the principle of a machine for laying cordage, which he sold to me in Paris as his own invention." A few years later, Cutting said, he met a man named Cartwright in England who had patented a similar machine there and whom Fulton knew. The implication was that Fulton had stolen Cartwright's idea and presented it as his own.

At this point, Fulton rose, red-faced, and asked to speak. Ogden initially objected but finally gave Fulton the floor. Fulton angrily denied everything Cutting said and ordered his lawyers to sue Cutting and Thornton for libel. He explained that he had discussed the device with Cartwright, who knew he had taken it to France to obtain a patent of importation, a perfectly legal act. He said he never claimed it as an original invention. He failed, however, to address a crucial question: Why was the term of his French rope-making machine patent for fifteen years, not ten, the term for imported patents?[46]

Furthermore, Fulton said, he never saw Fitch's plans while in Paris, although he admitted knowing Vail while there. To prove he was first to think of putting paddlewheels on steamboats, he brought out what he described as his original draft of the letter he had sent to the earl of Stanhope in England in November 1793. In it he reported on his experiments with steamboats, mentioning that side paddlewheels were part of his plan. Ogden asked to see the letter, and Fulton handed it him. He held it to the light for a few seconds, then handed it back to Fulton and said, with a smile, that the paper it was written on contained an American watermark. Fulton shot back that he meant to say it was a true copy of the original draft, which was so old and torn that he had destroyed it after making this copy. A rush of whispers ran through the galleries.[47]

Ogden's attorney, Hopkinson, later asked Fulton why he had never taken a single patent infringer to court; there had certainly been a few. At this, Fulton "exploded." He said the patent was not the issue, that the monopoly was the law in force in this case, and that if Ogden continued to run his steamboat on the Hudson, Fulton would seize it, as the New York law allowed—and not only that, he wouldn't hesitate to shoot Ogden.[48]

Hopkinson, on the final day of the weeklong hearings, noted that Fulton's achievements as an entrepreneur were worthy of praise, but that he was not a true inventor. Rather, he was a borrower whose wealth came from government patronage. He lashed out at Fulton for having someone else sign his patent application, and he hinted at perjury in the Stanhope letter incident. Leaving the impression that Fulton was a liar and a cheat, he wrapped up his case.

Emmet, Fulton's lawyer, barely knew where to start. "His eyes seemed to flash like lightning, and his manner and attitude were in a high degree dignified and commanding. He was indignantly agitated, and his gesture and appearance reminded me of an old ocean vexed by storms," wrote one of the eyewitnesses.[49] Emmet vigorously attacked Thornton, criticizing his abuse of office. At one point, as Fairfax reported in a letter to Thornton, "he became so vexed and angry that he forgot his argument; and having spit his spite, became rather vapid."[50] Fairfax added that it was "a proud day for the ghost of old John Fitch and one which the man himself would have highly enjoyed."

A few days later, the legislature voted along party lines in favor of Fulton and Livingston. Even though Ogden had recently served as New Jersey governor, he had little influence; he was a federalist, while the legislature was mostly republican. The real issue, interstate commerce, was left for another day and another venue, the U.S. Supreme Court. When the time came, years later, Ogden would again play a leading role.

Although Fulton won the latest battle in the steamboat wars, his victory must have been bittersweet given the attacks on his character. He and his attorneys packed up and left Trenton. As they approached Manhattan by ferry, they hit thick ice. They climbed out of the boat and began walking the rest of the way to the shore on the frozen river. Emmet, a heavy man, went through the ice at a thin spot, slipping into the frigid water. Fulton grabbed Emmett's arm and was able to pull him to safety, drenching himself in the process. He walked the rest of the way home, cold and wet, and took to his bed for days. On the morning of February 23, he died at the age of forty-nine, of pneumonia complicated by a lifetime of lung problems, probably tuberculosis.

Regardless of who he was or how he did it, Fulton, helped considerably by those who came before him, pushed America ahead. In less than eight years, he had put thirteen steamboats into service and was building three more. As George Washington had envisioned, faster, cheaper, and more reliable modes of transportation would fuel the nation's growth westward and keep the nation whole. The force of steam had been captured and proved, and it would power the nation's boats, ships, railroads, mills, and factories for more than a century to come.

EPILOGUE

The time will come when people will travel in stages moved by steam engines, from one city to another, almost as fast as birds fly, fifteen or twenty miles in an hour.

—*Oliver Evans*

The Fulton-Livingston steamboat heirs tried to enforce their monopolies on the Hudson and the Mississippi for several years after Fulton's death. But the fortunes to be made were too great. Steamboat companies continued to form anywhere there was a market. Thanks to Thornton's stinging criticism of Fulton's patents, no one much worried about infringement suits, and in fact the Fulton contingent never brought one.

The steamboat era on the Mississippi began in earnest when Daniel French and his captain, Henry Miller Shreve, launched the *Enterprise* on the Monongahela River in 1815. It was the first steamboat to make it back up the Mississippi from New Orleans to Louisville, a feat it accomplished in twenty-five days. Shreve was soon building his own boats, and his 1817 *Washington* is considered the first truly successful Western riverboat.[1] A couple of years later, tired of fighting off threats from Edward Livingston in New Orleans, Shreve challenged the Fulton-Livingston monopoly on the lower Mississippi in court and won. Shreve's improvements to Oliver Evans's high-pressure steam engines provided the model for the steamboats that followed.

Back east around that time, Aaron Ogden gave in and bought a license from the Fulton-Livingston monopoly so he could legally run his steam ferry on the Hudson. In 1819, Thomas Gibbons, a former business partner of Ogden's, decided that the New York State monopoly was unconstitutional. He obtained a federal coasting license and began running his steam ferry on the same routes as Ogden and John R. Livingston. Gibbons's captain was an ambitious young man named Cornelius Vanderbilt. He charged one dollar for tickets when the going rate was four. Naturally, the boat was loaded with people, who were so thrilled with their bargain fare that they hardly noticed the inflated prices they were paying for food and drink at the bar. Vanderbilt began building his own steamboats ten years later, and by the 1840s he owned more than a hundred.

Ogden, who never minded a good fight, asked the New York court of chancery to put a stop to Gibbons's boat. Ogden won round one. Gibbons appealed to the state court of errors, all the while continuing to operate his ferry. Ogden won again. Gibbons then appealed to the U.S. Supreme Court. In a landmark case in 1824, with Thomas Emmet again representing the Fulton-Livingston interests and Daniel Webster representing Gibbons, Chief Justice John Marshall ruled that the Constitution gave the federal government the power to control interstate commerce. With a pound of the gavel, the Fulton-Livingston monopoly was dead. By the late 1830s, more than a hundred steamboats were in service on the Hudson, and another 700 or so were carrying passengers and cargo on rivers, lakes, and bays all over the eastern United States.[2]

The steamboat pioneers passed into history. In a funeral attended by thousands of mourners, the new national hero Robert Fulton was buried in the Livingston family vault at Trinity Church in New York City. A few years later, Partition Street, a small road that connected the two Fulton ferry landings on the East River and the Hudson, was improved and renamed Fulton Street. The *North River* was retired a few years later, and its fate remains unknown. In textbooks forever after, it would be remembered as the *Clermont*. Harriet remarried not long after her husband's death, left their children with relatives, and moved to England.

Benjamin Latrobe, unemployed after Fulton fired him in 1814, was able to get his old job back as architect of the Capitol after the British burned it. While back in Washington, he helped design the distinctive north and south porticos of the White House, as well as St. John's Episcopal Church and the Stephen Decatur House on Lafayette Square. Asked by Jefferson to help with designs for the University of Virginia, he suggested the pavilion plan and the domed building that Jefferson later designed as the Rotunda.[3] Disputes with President Monroe led him to quit his Capitol architect job after two years. He moved to Baltimore, where he completed his design work for the Basilica of the Assumption and the Baltimore Exchange. In 1819, he moved to New Orleans to supervise the completion of its waterworks. He died a year later of yellow fever at the age of fifty-six, three years to the day after his beloved son Henry succumbed to the same disease there.[4]

Latrobe's son-in-law, Nicholas Roosevelt, continued to be involved in engineering projects in New York City for many years. He retired at age seventy-two and moved with Lydia to Skaneateles, New York, where he died in 1854. Lydia, ever tough, lived another twenty-four years.

John Stevens and his sons continued to build steamboats and steam ferries, but in 1812 he made the visionary leap, right over canals, to railroads. He wrote tirelessly to influential citizens, including DeWitt Clinton, Jefferson, Monroe, and Madison, describing a nationwide network of railroads that could connect towns and cities everywhere, not just those along waterways.[5] In 1813, he tried to convince Erie Canal supporters that his steam railway would be cheaper to build and could operate even in the winter. He predicted, "I can see nothing to hinder a steam-carriage from moving with a velocity of one hundred miles an hour." That last statement convinced them the old fellow was senile, and they brushed him aside.[6]

Three years later, in 1815, the state of New Jersey granted Stevens the first railroad charter in America, for a rail line to run between New Brunswick and Trenton. It went nowhere because Stevens was unable to attract financial backers for such a loony idea. Around 1825, he built a small steam locomotive that he ran on a circular track at his estate in Hoboken, just to show it could be done.[7] In 1830, a second state charter to Stevens made the Camden & Amboy Railroad a reality. Operated by his sons Edwin and Robert (John was well into his eighties by then), the first railway line in the United States began taking passengers three years later. (The Baltimore & Ohio railroad ran its first steam locomotive around that time as well.) The train was powered by a ten-ton steam locomotive called the John Bull, which Robert Stevens ordered from Robert Stephenson and Company, the English railroad pioneers, and had it shipped to New Jersey in pieces.[8] The original John Bull can be seen at the Smithsonian's National Museum of American History in Washington, D.C.

Oliver Evans, disgusted with the lack of fame and fortune his inventions had brought him, spent the last years of his life developing his Mars Works in Philadelphia, where he built steam engines, boilers, and other machines and machine parts. He also helped his son, George, build the Pittsburgh Steam Engine Company. Most of the steam engines used in Mississippi and Ohio River steamboats were modeled after the Evans high-pressure engine, a much more powerful and lighter-weight engine than any Watt-type engine.[9] As a comparison, consider that Fulton's *North River* Boulton & Watt steam engine generated about twenty-eight horsepower, about the size of an average riding lawn mower today. Shreve's *Washington,* with its high-pressure Evans-type engine, operated at about one hundred horsepower.[10]

Evans was probably the most inventive genius of his time. He once made a list of his eighty inventions, a variety of items that included refrigeration machines, a bread-kneading machine, a plan for gas lighting, and a machine gun. The Mars Works burned to the ground in 1819, four days before he died. His grist mill designs set the standard in America and eventually throughout Europe. He was also the first inventor to test the strength of federal patents in the courts. He took dozens of millers to court when they refused to pay him royalties, enforcing the idea that intellectual property was as valuable as tangible property.[11]

William Thornton never did build a steamboat, but he did design a few more homes and buildings, including Pavilion VII at the University of Virginia, after receiving a request from Jefferson. He lobbied hard to get a diplomatic posting to one of the newly independent South American countries, but his sometimes-dubious reputation as a government employee made him seem unsuitable for such a job. He remained at the Patent Office until his death in 1828. His funeral was a grand affair, though, and was attended by President John Quincy Adams, members of Congress, the cabinet, and other Washington dignitaries. He was buried in Congressional Cemetery, his grave marked by a cenotaph designed for that cemetery by none other than his nemesis, Benjamin Latrobe.[12] Given a chance, Thornton surely would have complained.

James Rumsey's remains rest in London at St. Margaret's Church, next door to Westminster Abbey, his achievements carved on a plaque in the church that was placed there in 1955 by the citizens of Shepherdstown, West Virginia. In 1915, an earlier group of townspeople erected a seventy-five-foot granite column to his memory in that town. It stands on a bluff overlooking the spot on the Potomac River where he ran his first steamboat. The Rumseian Society is alive and well, and in 1987 its members built a half-scale working model of Rumsey's first steamboat, engine and all. Every so often, the group sponsors the Rumsey Regatta, a town-wide celebration in which the members take their steamboat out of its shed behind the town museum, put it into the river, fire up the boiler, and run it up and down the Potomac for several hours.

In the early 1900s, sparked by centennial celebrations of Fulton's first trip up the Hudson, the steamboat wars resumed between the Fitch and Rumsey contingents. Several pamphlets and magazine articles, mostly originating in West Virginia, brutally attacked Fitch's character and claims, using facts and figures that weren't always correct or complete.[13] In March 1930, a Fitch descendant sent a letter to the *New York Times* demanding that Fitch's name stand with Fulton's in a hall of fame. For nearly a month, Rumsey supporters responded with long letters defending their inventor.[14] Rumsey's biographer, Ella May Turner, put an end to the letter wars by rightly stating that all three men should be credited with steamboat achievements.

As for Fitch, in 1926, the U.S. Congress approved funds for a granite monument to be erected in his honor across from the courthouse in Bardstown, Kentucky, where he died in 1798. His remains were moved there from the town cemetery the following year. Plaques or monuments to him exist at the capitol in Hartford, Connecticut; on the lawn of a bank at the corner of York and Street Roads in Warminster, Pennsylvania, where the idea for a steam carriage first hit him; and a spot near the Delaware River in Trenton. The most significant recognition of his achievements is the 1876 fresco by Constantino Brumidi showing Fitch (actually, a best guess of his likeness, since no life portrait was ever made of him) working on a model steamboat. It is located in the Senate wing of the U.S. Capitol, in an ornate hallway where the committee on patents was once located. It sits above a doorway, facing frescoes of the only other inventors recognized there—Benjamin Franklin and Robert Fulton.

Undoubtedly the most intriguing tribute to Fitch is a model that he crafted in Bardstown a year or two before he died. Described earlier, it is some kind of steam vehicle featuring four flanged wheels. It sat in his daughter's attic in Worthington, Ohio, for half a century after his death. A granddaughter remembered seeing it there as a child, and her son rediscovered it in 1849. It was loaned to the St. Louis Mercantile Library in 1854, where it was displayed for several years until being returned to the family for safekeeping during the Civil War.

From the beginning, there was much head scratching as to the purpose of the machine the model depicts. The St. Louis librarians decided it must have been a plan for a steamboat that would run on submerged tracks, a theory later discounted as "absurd on the face of it."[15]

In 1953, the curator of the land transportation section of the Smithsonian Institution heard about the model and began a search to locate it, finally tracking it down at the Ohio State Museum. It had been sitting there since 1903, a donation from the Fitch great-grandson who had discovered it.[16] On receiving the letter from the Smithsonian, the curator dusted the model off, had an expert study it, and concluded, "I am more firmly convinced than ever that Fitch actually constructed the prototype of a practical land-operating steam machine. . . . Fitch embodied in his engine the principle of the full power stroke, eliminating the loss of power on the upward motion of the piston. The engine is so constructed that it can be made to travel either forward or backward."[17]

The model features a boiler design that would become widely used years later. "Hauntingly, this [model] has a firebox cylinder within a cylindrical boiler," noted Brooke Hindle, former historian emeritus and director of the National Museum of American History. This design was first tried by Oliver Evans in the late 1790s and was later known as a Cornish boiler, after the type Richard Trevithick used in his steam locomotive engines in Britain after 1812.[18]

Today Fitch's model is on permanent display at the Ohio Historical Center in Columbus. The label that describes it ends with these words: " . . . most agree it is the world's first self-propelled railway steam locomotive."

CHRONOLOGY

1784

September: James Rumsey demonstrates mechanical boat model to George Washington in Bath, Virginia.

November: Rumsey tells Washington of his idea to use steam to power a boat.

1785

April: John Fitch has idea for a steam carriage, then a steamboat.

July: Rumsey is named superintendent of the Potomac Navigation Company.

September: Fitch presents steamboat plans to the American Philosophical Society.

December: Rumsey and Barnes install steam machinery on a boat near Harpers Ferry, Virginia; weather prevents a test.

1786

April: Fitch moves to Philadelphia, forms the Steamboat Company.

Spring: Robert Fulton visits Bath, Virginia.

July: Rumsey quits the Potomac Navigation Company.

September: Rumsey tests the mechanical poleboat, unsuccessfully.

December: Rumsey tests the pipe boiler, unsuccessfully.

1787

Spring: Fulton has moved from Philadelphia to London.

May: Fitch launches the *Perseverance*, using a rack of paddles over the sides.

August: Fitch demonstrates the *Perseverance* to delegates of the Constitutional Convention.

December: Rumsey demonstrates water-jet-propelled steamboat in Shepherdstown, Virginia.

1788

March: Rumsey comes to Philadelphia, publishes pamphlet aimed at Fitch.

May: Fitch publishes counter-pamphlet; Rumsey departs for England.

July: Rumsey attempts to form a partnership with Boulton & Watt.

October: Fitch's redesigned steamboat makes several trips to Burlington.

November: Rumsey receives English patent for his steamboat.

1789

March: Rumsey meets Jefferson in Paris.

Winter: Rumsey makes unsuccessful test of the *Columbian Maid* in London.

1790

April: Patent Act of 1790 signed into law.

June: Fitch begins steamboat passenger service between Philadelphia and Trenton.

1791

March: Rumsey signs on new partners in London.

April: Jefferson awards simultaneous patents to steamboat inventors.

1792

Summer: Fitch continues to work on steamboat in Philadelphia; receives French patent for steamboat.

December: Rumsey dies in London.

1793

January or February: The *Columbian Maid* makes first and only trial in London.

February: Fitch leaves for France to build steamboat.

December: Fitch leaves France, arrives in London.

1794

Spring: Fitch returns to United States.

1796

Fitch moves to Kentucky.

1797

Spring: Fulton moves to Paris.

1798

July: Fitch dies in Kentucky.

1802

March: Fulton meets Livingston in Paris.

1803

August: Fulton tests steamboat on the Seine.

1804

May: John Stevens tests *Little Juliana,* first screw-propeller steamboat, in New Jersey.

1805

July: Oliver Evans runs steam-powered amphibious vehicle in Philadelphia.

1806

December: Fulton returns to the United States.

1807

August: Fulton runs *North River* steamboat from New York City to Albany.

1811

October: Fulton's *New Orleans,* the first steamboat on the Ohio and Mississippi, departs Pittsburgh.

1815

February: Fulton dies in New York.

NOTES

In citing works in the notes, usually short titles have been used. Works cited frequently are identified by the following abbreviations.

APS	American Philosophical Society Library
FP	Fitch Papers, Library of Congress
HSP	Historical Society of Pennsylvania
LCP	Library Company of Philadelphia
LOC	Library of Congress
GW Papers	George Washington Papers, Library of Congress, American Memory website, online at http://memory.loc.gov/ammem/gwhtml/gwhome.html
TJ Papers	Thomas Jefferson Papers, Library of Congress, American Memory website, online at http://memory.loc.gov/ammem/mtjhtml/mtjhome.html
NMAH	National Museum of American History, Transportation Archives
RS	Rumseian Society
VHS	Virginia Historical Society

Full citations may be found in the bibliography.

Notes to Chapter 1: Stream-boats and Steamboats

1. Hulbert, *Washington and the West*, 32.
2. Jackson and Twohig, *Diaries of George Washington*, vol. 2, note to diary entry for 23 August 1769.
3. Ibid., note 1 to diary entry for 6 September 1784.
4. LOC, GW Papers, George Washington to James Craik, 10 July 1784.
5. Hulbert, *Washington and the West*, 125, quoting Sparks, *Writings of Washington*, vol. 9, 105.
6. Ibid., 32.
7. Brooke Hindle writes (in *Pursuit of Science in Revolutionary America*, 374) that "Abner Cloud invented a similar boat in Pennsylvania, [and] John Mason conceived still another sort of mechanical boat."

8. Jackson and Twohig, *Diaries of George Washington,* vol. 4, note 2 to diary entry for 6 September 1784.

9. Hulbert, *Washington and the West,* 21, 37.

10. Freeman, *George Washington: A Biography,* 14.

11. Fitzpatrick, *Writings of George Washington,* letterbook, note 70, 24 June 1784.

12. Hulbert, *Washington and the West,* 11–12.

13. Jackson and Twohig, *Diaries of George Washington,* vol. 4, diary entry for 3 September 1784.

14. Hulbert, *Washington and the West,* 17.

15. Ibid., 20, 124.

16. LOC, GW Papers, GW to Francois Jean, Comte de Chastellux, 12 October 1783.

17. LOC, GW papers, Thomas Jefferson to GW, 18 March 1784.

18. Jackson and Twohig, *The Diaries of George Washington,* vol. 4, diary entry for 6 September 1784.

19. Turner, *James Rumsey,* 2, 67.

20. Ibid., 6–7.

21. Ibid., 4–6.

22. LOC, GW papers, GW to Rumsey, 7 September 1784.

23. LOC, *Journals of the Continental Congress,* 11 July 1783.

24. LOC, GW Papers, Rumsey to GW, 10 March 1785.

25. Hulbert, *Washington and the West,* 103.

26. LOC, GW Papers, GW to Benjamin Harrison, 10 October 1784; Harrison to GW, 13 November 1784; GW to Benjamin Harrison, 10 October 1784, note 56.

27. Turner, *James Rumsey,* 15, 65.

28. LOC, TJ Papers, Madison to TJ, 9 January 1785.

29. The tremendous difficulties in achieving this goal eventually became apparent—for example, the seventy-six-foot drop in river level at Great Falls required building five locks, some of the first in the nation. Work continued off and on until 1802, when the system finally opened for traffic. But low water levels limited transport to just a few months each year. In 1821, Virginia and Maryland revoked the charter to the Potomac Company and later transferred rights to the Chesapeake and Ohio (C&O) Canal Company in 1828. (National Park Service, Historic American Engineering Record, "Potomack Company: Great Falls Canal and Locks, Great Falls, Fairfax County, Virginia," HAER No. VA - 13, 1982.) Note: Modern spellings are used throughout this book; here, "Potomac" was spelled several ways for many years, most commonly "Potowmack."

30. Freeman, *George Washington,* 26–27; Hunter, "James Rumsey," 186.

31. Turner, 64, 87.

32. Williams, *History of Invention,* 172.

33. Eco and Zorzoli, *Pictorial History of Inventions,* 195.

34. Ibid.

35. Bruno, *Tradition of Technology,* 137–38.

36. Pursell, *Early Stationary Steam Engines,* 6.

37. Pursell, *The Machine in America,* 57.

38. York, *Mechanical Metamorphosis,* 53. Colles later became involved in canal schemes in New York State and created the first American road maps.

39. Pursell, *Early Stationary Steam Engines,* 9; "Statewide Historical Preservation Report P-C - 1" (September 1980): 69–70. Providence: Rhode Island Historical Preservation Commission, 1980.

40. Cruse, "The Projector Detected."

41. Curtis, "Rumseian Experiment," 21.

42. McFarlane, *History of Propellers,* 15.

43. Trevor, *History of Invention,* 172.

44. *Journals of the Continental Congress,* 8 December 1784.

45. Ibid., 11 May 1785; *Letters of Delegates to Congress,* Hugh Williamson to TJ, 11 December 1784.

46. LOC, GW Papers, GW to Hugh Williamson, 19 March 1785.

47. LOC, *Journals of the Continental Congress,* 11 May 1785.

48. LOC, GW Papers, Rumsey to GW, 24 June 1785.

49. LOC, GW Papers, GW to Rumsey, 2 July 1785.

50. Turner, *James Rumsey,* 39.

51. Ibid., 66–69.

52. Ibid., 71.

53. Turner, *James Rumsey,* 73.

54. LOC, GW Papers, GW to Rumsey, 31 January 1786.

Notes to Chapter 2: A Ridiculous Idea

1. Jackson and Twohig, *Diaries of George Washington,* entry for 4 November 1785.

2. Watson, *Annals of Philadelphia and Pennsylvania,* ch. 90.

3. Prager, *Autobiography,* 156–57.

4. Ibid.

5. York, *Mechanical Metamorphosis,* 89–91.

6. LOC, GW Papers, Series 5, Revolutionary War Accounts, Vouchers, and Receipts, Fitch to Clear Gibbs, 26 November 1778. In addition to his personal land purchases over the years, Washington was himself involved in a land speculation scheme after the French and Indian War. He and several other Virginians organized a group called the Mississippi Company. They applied to England for a land grant for some 2.5 million acres in the Mississippi valley, but the grant was never made. (Friedenberg, *Life, Liberty and the Pursuit of Land,* 112.)

7. LOC, GW Papers, GW to James Duane, 7 September 1783.

8. LOC, GW papers, Rumsey to GW, 24 June 1785. Rumsey never completed Washington's large house.

9. Because of the switch from the Julian to the Gregorian calendar in 1752, Fitch's original birth date of January 21, 1743, was pushed forward by eleven days and one year. The change in year was necessary because under the Julian calendar, Britain and her colonies started the new year on March 25, not January 1.

10. Frank D. Prager's *Autobiography of John Fitch* is the edited version of Fitch's "Life" and the autobiographical portions of his "Steamboat History," both of which Fitch deposited with the Library Company of Philadelphia in 1792. These materials and other Fitch documents and correspondence comprise the Fitch Papers at the Library of Congress. Most of the information about Fitch's life and endeavors contained in this chapter and elsewhere in this book were drawn from this source.

11. Copies of both books can be found in the Library of Congress. *Hodder's Arithmetick,* by James Hodder, was published in Boston in 1719; *A New Geographical and Historical Grammar and the Present State of the World,* by Thomas Salmon, was published in London in 1758.

12. LOC, *Journals of the Continental Congress,* 27 January 1785.

13. During that period he boarded at a house owned by Charles Garrison, which still stands, according to the Warminster, Pennsylvania, website: www.warminstertownship.org.

14. Prager, *Autobiography,* 113.

15. Gilbert, "Oliver Evans' Memoir," 152.

16. This church, one of the oldest Presbyterian churches in America, is now called the Neshaminy Warwick Presbyterian Church. It was founded in 1727, and Nathaniel Irwin was the pastor from 1774 until 1812.

17. This was Benjamin Martin's *Philosophia Britannica; or, A new & comprehensive system of the Newtonian philosophy, astronomy & geography. In a course of twelve lectures, with notes, containing the physical, mechanical geometrical & experimental proofs & illustrations of all the principal propositions in every branch of natural science. Also a particular account of the invention . . . of all the considerable instruments, engines, & machines.* Various editions were published in England throughout the 1700s.

18. Many historians think Fitch was not being quite truthful. For one thing, the Schulyer Newcomen engine in Belleville, New Jersey, was somewhat of a tourist attraction, even after it fell into ruins, and Fitch had lived in and traveled around New Jersey for several years before and during the Revolution. Pursell, *Early Stationary Steam Engines,* 6.

19. Watson, *Annals of Philadelphia and Pennsylvania,* ch. 90.

20. Prager, *Autobiography,* 145.

21. LOC, *Journals of the Continental Congress,* 30 August 1785.

22. Ibid., 1 September 1785; Smith, *Letters of Delegates to Congress*, vol. 23, Josiah Hornblower to John Fitch, 24 May 1786, note.
23. Prager, *Autobiography*, 153.
24. LOC, FP, Fitch to APS, 1 August 1785.
25. Prager, *Autobiography*, 154.
26. LOC, TJ Papers, David Rittenhouse to TJ, 14 April 1787.
27. Conway, *Life of Thomas Paine, 280–81.*
28. Jordan, Francis, Jr., *Life of William Henry*, 53; Prager, *Autobiography*, 155.
29. Prager, *Autobiography*, 156.
30. Ibid.
31. LOC, FP, certificate, City of Richmond, 16 November 1785.
32. Prager, *Autobiography*, 160.
33. Ibid., 166.
34. Franklin had met Matthew Boulton when he was living in England, and in 1758 they began a lifelong correspondence on scientific subjects including electricity, metallurgy, and steam power.

Notes to Chapter 3: Brother Saintmakers

1. Turner, *James Rumsey*, 39.
2. Ibid., 51. All following information about Rumsey is from this source.
3. LOC, *Diaries of George Washington*, entry for 3 October 1786; LOC, GW Papers, Rumsey to GW, 19 September 1786.
4. Sale, *Fire of His Genius*, 46; Philip, *Robert Fulton*, 11–12.
5. LOC, GW Papers, Rumsey to GW, 19 September 1786.
6. Ibid.
7. Turner, *James Rumsey*, 76.
8. York, *Mechanical Metamorphosis*, 49, 185; Prager, *Autobiography*, 162–63; Prager, "Steamboat Pioneers," 508–10.
9. Prager, *Autobiography*, 166–67.
10. Ibid., 163.
11. LCP, John Hall Diary, entry for 10 March 1786.
12. Green, "Report," 8–12.
13. The elder Hall is mistaken for his nephew in an article in the Newcomen Society *Transactions*, "Thomas Paine's Iron Bridge Work," 190–91.
14. APS, *Transactions*, 2 (1786): 308–309.
15. APS, Franklin Papers, BF to Francis Hopkinson, 27 March 1786.
16. Prager, *Autobiography*, 166.
17. Ibid., 167.
18. LOC, FP, Fitch to Colles and Hornblower, 17 May 1786.
19. LOC, FP, Hornblower to Fitch, 24 May 1786.
20. Bedini, "Henry Voigt."
21. Dickinson and Jenkins, *James Watt*, 23.

22. Uglow, *Lunar Men,* 244.
23. Ibid., 59.
24. Hindle, *Emulation and Invention,* 72.
25. LOC, FP, Fitch to Franklin, 4 September 1786.
26. *Columbian Magazine,* December 1786, 174.
27. *Graham's Magazine,* 1842, 108.
28. Prager, *Autobiography,* 179.
29. Boyd, *Papers of Thomas Jefferson,* vol. 11, 561, Francis Hopkinson to TJ, 8 July 1787.
30. LOC, GW Papers, Thomas Johnson to GW, 16 November 1787.
31. LOC, FP, several items of correspondence between Fitch and St. John de Crèvecoeur over the next few years.
32. APS, Franklin Papers, St. John de Crèvecoeur to BF, 30 January 1788; FP, statements of Rittenhouse and Ewing, 12 December 1787, and of Andrew Ellicott, 13 December 1787.
33. APS, Franklin Papers, BF to St. John de Crèvecoeur, 16 February 1788.
34. APS, Franklin Papers, BF to Jean-Baptiste Le Roy, 17 February 1788.
35. Prager, *Autobiography,* 181.

NOTES TO CHAPTER 4: THE WAR OF THE PAMPHLETS

1. Turner, *James Rumsey,* 76. See also this source for the information that follows on Rumsey's first trials and the events surrounding them.
2. Rumsey, "A Short Treatise on the Application of Steam," certificate no. 11, Philadelphia, 1788.
3. LOC, GW Papers, Rumsey to GW, 17 December 1787.
4. The Rumseian Society of Shepherdstown, West Virginia, occasionally demonstrates its one-half-scale working model of Rumsey's steamboat on the Potomac River below the town; the sounds described were heard there by the author in July 2000.
5. LOC, GW Papers, GW to Thomas Johnson, 22 November 1787.
6. LOC, GW Papers, Rumsey to GW, 17 December 1787.
7. Turner, *James Rumsey,* 111; Rumsey, "Short Treatise."
8. LOC, GW Papers, Rumsey to GW, 24 March 1788.
9. Cruse, "The Projector Detected," Baltimore, 1788.
10. Turner, *James Rumsey,* 118.
11. Rumsey, "Explanation of a Steam Engine," Philadelphia, 1788.
12. Turner, *James Rumsey,* 109; Barnes, "Remarks on Mr. John Fitch's Reply."
13. APS, Rumsey Papers, John Vaughan to Benjamin Vaughan, 15 May 1788.
14. LOC, FP, letter dated 14 April 1788; Boyd, Papers of Thomas Jefferson, vol. 15, 641–42. The footnote to this letter notes that it was not recorded in the SJL Index, that SJL sheets for 1788 are missing, that no other copy exists of this letter or a reply to it, and that it "is possible that Fitch did not dispatch it" [unlikely] "or that TJ did not receive it."

15. Wells had been presenting his ideas to APS members since the 1770s, including plans for a wind-operated ship's pump, a release spring for carriages, and a steam cannon (York, Mechanical Metamorphosis, 51).

16. Ibid., 181–82.

17. Fitch, "The Original Steamboat Supported."

18. Prager, *Autobiography*, 157.

19. Prager, *Autobiography*, 186; Fitch, "The Original Steamboat Supported."

20. Barnes, "Remarks on Mr. John Fitch's Reply."

21. Fitch, "The Original Steamboat Supported."

22. Turner, *James Rumsey*, 121–22; also LOC, Rumsey Papers

23. VHS, Rumsey Papers.

24. LOC, GW Papers, Rumsey to GW, 15 May 1788.

25. New York Assembly Papers, Miscellaneous, vol. 3.1. Petition of the Rumseian Society.

26. Barnes, "Remarks on Mr. John Fitch's Reply."

27. *Journals of the Continental Congress*, 11 February 1788, 5 March 1788.

28. LOC, FP, "Steamboat History," 57–58.

29. Prager, *Autobiography*, 183.

30. Westcott, *Life of John Fitch*, 249.

31. Ibid., 251–52.

32. Prager, *Autobiography*, 183.

33. Ibid., 184.

34. Harris, *Papers of William Thornton*, xxxiii–xxxv; Stearns and Yerkes, *William Thornton*, 4, 19–20.

35. Prager, Autobiography, 187; LOC, FP, Fitch to John Hall, n.d. December 1788; Fitch to Steamboat Company, 5 December 1788.

36. Prager, *Autobiography*, 185.

37. Ibid.

Notes to Chapter 5: The Columbian Maid

1. Prager, *Autobiography*, 124–25.

2. Ibid.

3. Ibid., 126.

4. Ibid., 127.

5. HSP, George Bryan Papers, box 3, f. 9.

6. HSP, Fitch Papers.

7. Westcott, *Life of John Fitch*, 265–66.

8. LOC, FP, Fitch letter to editor, 28 March 1789.

9. Prager, *Autobiography*, 186.

10. LOC, FP, Fitch's undated account of incident.

11. Thornton, "Short Account."

12. Turner, *James Rumsey*, 141.

13. HSP, Simon Gratz Collection, Rumsey to William Bingham, 3 August 1788. Bingham was a wealthy Philadelphia merchant, banker, and politician who was heavily involved in land speculation around the time he supported Rumsey.

14. Turner, *James Rumsey,* 143.

15. Hindle, *Emulation,* 40.

16. Boyd, *Papers of Thomas Jefferson,* vol. 13, 346, Vaughan to TJ, 11 July 1788.

17. APS, Rumsey Papers, TJ to Benjamin Vaughan, 23 July 1788.

18. HSP, Simon Gratz Collection, Rumsey to Boulton & Watt, 6 August 1788.

19. Turner, *James Rumsey,* 152.

20. RS, Matthew Boulton to Rumsey, 29 August 1788.

21. Turner, *James Rumsey,* 150–51.

22. Ibid., 154.

23. Ibid., 156.

24. Ibid., 157. Dorothy Jordon was a famous London actress of the day; she was also the long-time mistress of William, the third son of King George III who later became King William IV; she bore him ten children.

25. Boyd, *Papers of Thomas Jefferson,* vol. 14, John Trumbull to TJ, 11 March 1789.

26. Ibid., Thomas Paine to TJ, 16 February 1789.

27. LOC, Rumsey Papers, Rumsey to Charles Morrow, 28 March 1789.

28. Ibid.

29. Woodress, *Yankee's Odyssey,* 90–94.

30. Belote, *Scioto Speculation.*

31. "Gallipolis."

32. LOC, Rumsey Papers, Rumsey to George West, 20 March 1789.

33. LOC, Jefferson Papers, Rumsey to TJ, 22 May 1789.

34. Boyd, *Papers of Thomas Jefferson,* vol. 15, 170–71, Rumsey to TJ, 6 June 1789.

35. RS, Rumsey to Charles Morrow, 4 August 1789; also LOC, Rumsey Papers, *McMechen* vs. *Rumsey.*

36. Boyd, *Papers of Thomas Jefferson,* 403–404, Rumsey to TJ, 8 September 1789.

37. Ibid.

38. Ibid., 413, TJ to Rumsey, 10 September 1789.

39. LOC, Jefferson Papers, Rumsey to TJ, 22 September 1789.

40. Boyd, *Papers of Thomas Jefferson,* vol. 15, 413, TJ to Rumsey, 10 September 1789.

41. LOC, Jefferson Papers, TJ to William Short, 27 April 1790.

42. LOC, Rumsey Papers, Rumsey to Charles Morrow, 7 February 1790.

43. Ibid.

Notes to Chapter 6: Lord High Admirals of the Delaware

1. Prager, *Autobiography,* 191.

2. Ibid., 191.
3. Bathe, *Engineer's Miscellany*, 48; LOC, FP, "Steamboat History," 109–110.
4. Ibid.
5. Westcott, *Life of John Fitch*, 282.
6. Boyd, *Poor John Fitch*, 226.
7. Thornton, "A Short Account," 5.
8. Westcott, *Life of John Fitch*, 287.
9. LOC, *The New York Magazine*, August 1790, 493.
10. Westcott, *Life of John Fitch*, 304–305. Oliver Pollock, interestingly, had once had a business relationship with a man named James Rumsey in 1775 that involved slave trading in the South, according to letters in the Rumseian Society files obtained from the Peter Force Collection at the Library of Congress. Whether this was the inventor James Rumsey is not certain, although Rumsey did own several slaves, a practice not common among the mostly German and Swiss settlers in western Virginia at the time.
11. LOC, FP, Fitch to Henry Knox, 9 March 1791.
12. LOC, FP, Fitch to William Samuel Johnson, 26 February 1790. Johnson, who had praised Fitch's first steamboat during the Constitutional Convention, was at the time serving as a U.S. senator from Connecticut. He would resign the following year to serve as full-time president of Columbia College (formerly Kings College), now Columbia University.
13. LOC, FP, Fitch to Henry Knox, 9 March 1791.
14. LOC, FP, copy of ad, March 1791.
15. Westcott, *Life of John Fitch*, 307–308.
16. Prager, *Autobiography*, 127.
17. Ibid., 127–28.
18. LOC, FP, Mary Krafft to Fitch, 18 March 1791.
19. Prager, *Autobiography*, 136–37.
20. Ibid., 128. Around this time, Fitch was disappointed in a second area of his personal life. He had formed a new religious group, which he called the Universal Society. Its tenets were similar to Unitarianism in that he believed in a "God of nature." Its beginnings were informal; he recalled how he and Voigt, admittedly a little inebriated ("we [were] frequently getting mildly glad in liquor"), would freely express their opinions about religion while working on their boat at the waterfront. They were surprised to find that many of the men they spoke with had similar beliefs. Like many educated people of the day, they called themselves deists, believing that while Jesus may have been a great man, he was not divine; they also rejected the concept of original sin. That March, Fitch, who had been trying to find a minister for the group, convinced a religious leader of a similar group, Elihu Palmer, to speak to a combined meeting of the Universal Society members and his own followers. The meeting went so well that Palmer took out a newspaper ad stating that he would speak again on the following Sunday. Palmer's boldness enraged the local

religious leaders, and a Bishop White "frightened" the owner of the hall into canceling the event. After that incident, no one in town would rent Palmer a hall. Fitch's short-lived hope that Palmer would lead his group was dashed, and his Universal Society slowly dissolved.

21. Westcott, *Life of John Fitch*, 320–24.

NOTES TO CHAPTER 7: THE FIRST AND TRUE INVENTOR

1. Walterscheid, "Defining the Patent and Copyright Term," 352, note 144.
2. The Statute of Anne was the world's first government law to protect the rights of authors and their literary creations, rather than printers and booksellers. It directly influenced early U.S. copyright law in that it granted limited terms and required registration and deposit of the book with a government office, along with a statement of copyright notice in the book.
3. Davis, *Decisions of the United States Courts*, x.
4. *Journals of the Continental Congress*, vol. 24, 2 May 1783, 326.
5. Unger, *Noah Webster*, 108.
6. Prager, "Steamboat Pioneers," 517–18.
7. Walterscheid, *Nature of the Intellectual Property Clause*, 96.
8. Prager, *Autobiography*, 179.
9. Prager, "The Steamboat Pioneers," 518.
10. Prager, "Historic Background," 309–10.
11. LOC, TJ Papers, TJ to Isaac McPherson, 13 August 1813.
12. Pursell, ed., *Technology in America*, 28.
13. LOC, TJ Papers, TJ to James Madison, 31 July 1788.
14. LOC, *Letters of Delegates to Congress*, vol. 25, James Madison to Thomas Jefferson, 17 October 1788. Jefferson also had to resolve his negative views on patents with the purpose of the constitutional clause that authorized them: "to promote the progress of science and useful arts." Jefferson had long held a bucolic vision of America as an agricultural society, preferring that the dirty and dehumanizing factories he saw in England remain on the other side of the Atlantic. But as an inventor himself he understood that technology could improve standards of living, and that the United States would need to keep pace with Europe if it were to grow and prosper. Jefferson eventually came to agree with Madison, albeit with reservations.
15. U.S. House Journal. 1st Cong., 1st sess., 20 April 1789.
16. Rollins, *Autobiographies of Noah Webster*, 235, 238; Prager, "Historic Background," 320.
17. Walterscheid, *To Promote the Progress*, 59–60.
18. Ibid., 84–87.
19. Ibid., 108, 173, note 82.
20. LOC, TJ Papers, Rumsey to TJ, 6 June 1789.

21. LOC, American Memory (http://memory.loc.gov), Annals of Congress, Senate, 1st Congress, 2nd Session, 969–70, president's address to both houses of Congress.
22. Walterscheid, *To Promote the Progress,* 175–78.
23. Boyd, *Papers of Thomas Jefferson,* vol. 27, application to the Patent Board, 10 May 1790.
24. Walterscheid, *To Promote the Progress,* 186.
25. *New American State Papers,* vol. 6, 11.
26. LOC, FP, Remsen to Fitch, 23 November 1790.
27. Flexner, *Steamboats Come True,* 147.
28. York, *Mechanical Metamorphosis,* 105.
29. Read, *Nathan Read,* 100.
30. Ibid., 111.
31. Ibid., 73.
32. Ibid., 53–54, 78–79.
33. Ibid., 93.
34. Boyd, *Papers of Thomas Jefferson,* vol. 27, 790–91.
35. Read, *Nathan Read,* 81; Hindle, *Emulation,* 45.
36. LOC, Isaac Briggs Papers; Pursell, *Early Stationary Steam,* 24.
37. LOC, FP, Remsen to Fitch, 5 March 1791.
38. Westcott, *Life of John Fitch,* 315.
39. Ibid.
40. LOC, FP, Memorial to Patent Board, 4 April 1791.
41. LOC, FP, Fitch to William Thornton, 27 April 1791.
42. Prager, "The Steam Boat Interference," 638.
43. LOC, FP, Fitch note to Steamboat Company after patent hearing, 25 April 1791.
44. LC, TJ Papers, TJ to Hugh Williamson, 1 April 1792. One example of his thoroughness is the 1791 patent application of Jacob Isaacs, who had devised a still and furnace to distill fresh water from sea water, an important advance in those days of long sea voyages. Jefferson asked Isaacs to bring his device and several gallons of seawater to his office, and then called in David Rittenhouse and another Philadelphia scientist to help judge its workings. After a week of testing, Jefferson declared Isaacs's device no better than existing ones and rejected his application.
45. LOC, FP, petition to the U.S. Congress, 15 February 1790.
46. Walterscheid, *To Promote the Progress,* 209. Barnes's pamphlet was entitled "Treatise on the Justice, Policy, and Utility of Establishing an Effectual System of Promoting the Progress of Useful Arts, by Assuring Property in the Products of Genius."
47. Ibid., 482.
48. LOC, TJ Papers, TJ to Isaac McPherson, 13 August 1813.

NOTES TO CHAPTER 8: "ALL FURTHER PROGRESS IS IN VAIN"

1. Harris, *Papers of William Thornton*, 140.
2. The elder Stockton was a New Jersey judge who, after signing the Declaration of Independence, was captured and harshly imprisoned by the British; he died, impoverished and in ill health, at the age of fifty-one in 1781.
3. Prager, *Autobiography*, 199.
4. Ibid., 200.
5. Harris, *Papers of William Thornton*, 142.
6. Ibid., 144.
7. Ibid., 155.
8. Ibid., xlvi-xlvii.
9. Prager, *Autobiography*, 201.
10. Harris, *Papers of William Thornton*, 156.
11. Ibid., 165.
12. Ibid., 231.
13. Boyd, *Papers of Thomas Jefferson*, vol. 27, 798, petition of Henry Voigt to the Patent Board, 10 August 1791.
14. Ibid., 133.
15. Ibid., 134.
16. Ibid., 135.
17. LOC, FP, statement signed by Henry Voigt, witnessed by George Krafft, 4 November 1791.
18. Westcott, *Life of John Fitch*, 333.
19. Harris, *Papers of William Thornton*, 177.
20. Prager, *Autobiography*, 3.
21. LOC, FP, Fitch to Mary Krafft, n.d. (addressed to her in New York).
22. Ibid.
23. LOC, FP, Mary Krafft to Fitch, n.d.
24. LOC, FP, Fitch to Mary Krafft, n.d.
25. Harris, *Papers of William Thornton*, 179.
26. LOC, FP, Fitch Steamboat Company accounts, 1 April 1792.
27. LOC, FP, Fitch to Dr. Benjamin Say and Edward Brooks, 6 May 1792.
28. HSP, Fitch Papers, 1786–1793.
29. Westcott, *Life of John Fitch*, 337; LOC, FP, Fitch to David Rittenhouse, 29 June 1792; HSP, Dreer Autograph Collection, Fitch to Richard Wells, 6 July 1792.
30. Westcott, *Life of John Fitch*, 338.
31. Ibid., 338–39.
32. Ibid., 341.
33. LOC, FP, Fitch to Hannah Levering, 27 September 1792.
34. LOC, FP, Fitch to John Israel, 23 October 1792.
35. Prager, *Autobiography*, 207.

36. Ibid., 208.
37. Bedini, "Henry Voigt."
38. Hindle, *David Rittenhouse,* 334.
39. LOC, FP, n.d. Fitch should have spelled "rhineo" as "rhino," an English slang word meaning cash or money. Although his talent for writing poetry is questionable, he surely meant "mineo" in the previous line to be pronounced "mine-o," not "min-e-o."
40. LOC, FP, Articles of Agreement between John Fitch and Elmer Cushing, 17 September 1792.
41. HSP, Dreer Autograph Collection, 50; LOC, FP, Agreement between John Fitch and John Nicholson, 4 January 1793.
42. HSP, Fitch Papers, 1786–1793.
43. Pennsylvania State Archives, Records of Nicholson Lands.
44. LOC, FP, Fitch power of attorney to William Thornton, 12 February 1793.
45. There is no evidence that Nicholson ever built a horseboat. He resigned as state comptroller in 1794 and partnered with the other leading Philadelphia financier of the time, Robert Morris, in several complicated land development deals. A few years later, both men ended up in debtors prison. The idea lingered on, however, and years later horse-powered boats were put to use as passenger ferries on the Hudson and East Rivers.

Notes to Chapter 9: Leeches and Sharks

1. LOC, Rumsey Papers, Rumsey to Charles Morrow, 7 February 1790.
2. Dabney, *Silver Sextant,* 137–38.
3. LOC, Rumsey Papers, Rumsey to Charles Morrow, 7 February 1790.
4. Ibid.
5. Boyd, *Papers of Thomas Jefferson,* vol. 14, John Trumbull to TJ, 11 March 1789.
6. LOC, GW Papers, GW to Daniel Parker, 12 September 1783.
7. Smith, *The Empress of China,* 63–64, 120.
8. LOC, Rumsey Papers, Rumsey to George West, 12 September 1791. Also in *William & Mary Quarterly,* "Letters of James Rumsey," April 1916, 240.
9. Turner, *James Rumsey,* 179.
10. *William & Mary Quarterly,* "Letters of James Rumsey," April 1916.
11. Turner, *James Rumsey,* 178.
12. NMAH, Rumsey files.
13. APS, Rumsey Papers, Rumsey to Levi Hollingsworth, 30 June 1790.
14. *William & Mary Quarterly,* "Letters of James Rumsey," April 1916, 245–46.
15. APS, Rumsey Papers, Rumsey to Levi Hollingsworth (Rumseian Society), 25–26 July 1791.
16. Ibid., 242–47.
17. "The Royal Canal," Inland Waterways Association.

18. *William & Mary Quarterly,* "Letters of James Rumsey," April 1916, 247. Also, APS, Rumsey Papers, Rumsey to Levi Hollingsworth (Rumseian Society), 25–26 July 1791.

19. *William & Mary Quarterly,* "Letters of James Rumsey," April 1916, 242, 274.

20. Ibid., 248.

21. Latyon, "Most Original," 56.

22. *William & Mary Quarterly,* "Letters of James Rumsey," April 1916, 244.

23. Ibid., 248.

24. LOC, TJ Papers, Thomas Jefferson to Charles Thomson, 22 April 1786; Pursell, *Early Stationary Steam Engines,* 15.

25. Turner, *James Rumsey,* 190.

26. LOC, Rumsey Papers, Rumsey to Charles Morrow, 30 March 1792.

27. LOC, Rumsey Papers, Rumsey to James McMechen, 15 April 1792; also in *William & Mary Quarterly,* "Letters of James Rumsey," January 1916, 171–72.

28. Turner, *James Rumsey,* 197–98.

29. Ibid., 198–99; LOC, Rumsey Papers, Claiborne to Miers Fisher, 2 January 1793.

30. Turner, *James Rumsey,* 198–99.

31. LOC, Jefferson Papers, John Brown Cutting to TJ, 24 December 1792.

32. RS, *Gentleman's Magazine,* London, February 1793, 182; Philadelphia *Aurora,* 13 April 1793.

33. LOC, Jefferson Papers, TJ to Le Roy, Pinckney, Morris, 24 May 1793.

34. Turner, *James Rumsey,* 201–202.

35. Ibid., 202–203.

36. Philip, *Robert Fulton,* 86.

37. Turner, *James Rumsey,* 209.

38. Flexner, *Steamboats Come True,* 170.

39. Pursell, *Early Stationary Steam Engines,* 24; Spratt, *Birth of the Steamboat,* 57–58.

40. Turnbull, *John Stevens,* 118–19.

Notes to Chapter 10: Mother Clay

1. LOC, FP, Fitch to William Thornton, 27 April 1793. On April 5, Thornton learned that his design for the U.S. Capitol had been approved by President Washington (Harris, *Papers of William Thornton,* 238).

2. LOC, FP, Fitch to William Thornton, 11 August 1793; Flexner, *Steamboats Come True,* 232.

3. LOC, FP, Fitch to William Thornton, 24 May 1793.

4. Paine, who had come to France in 1787 and whose influential pamphlets supporting the American Revolution were much admired there, had since

been made a French citizen and elected to the Convention. He managed to elude arrest for several months, but in December he was caught and thrown into the Luxembourg prison. He spent more than nine months in a dank cell, where he worked on his pamphlet, *The Age of Reason,* until James Monroe arrived as the U.S. minister to France and managed to get him released.

5. LOC, FP, Fitch to William Thornton, 11 August 1793.

6. Boyd, *Poor John Fitch,* 280. Johnson, whose brother was former Maryland Governor Thomas Johnson, was a partner in an American business that had operated in Europe for many years. Although he had lived primarily in London, he spent the years of the American Revolution in Nantes, France. He was named U.S. consul in 1790, and a few years later Johnson's only daughter, Louisa Catherine, married John Quincy Adams in London.

7. Ibid.

8. Ibid., 278.

9. Westcott, *Life of John Fitch,* 357–58.

10. LOC, FP, Fitch to William Thornton, 11 August 1793. Attempts to learn more about the utility of this chart have been unsuccessful.

11. LOC, FP, Fitch to William Thornton, 27 November 1793; Boyd, *Poor John Fitch,* 279.

12. Harris, *Papers of William Thornton,* 277.

13. Turner, *James Rumsey,* 141.

14. Ibid., 275.

15. LOC, FP, William Thornton to Fitch, 21 February 1794; Harris, *Papers of William Thornton,* 278.

16. Boyd, *Poor John Fitch,* 284.

17. Harris, *Papers of William Thornton,* 283n.

18. Ibid., 281, 285.

19. LOC, FP, Fitch to Thornton, 20 September 1794.

20. Harris, *Papers of William Thornton,* 286.

21. Turnbull, *John Stevens,* 127; Flexner, *Steamboats Come True,* 234.

22. Turnbull, *John Stevens,* 126.

23. Ibid., 127.

24. Westcott, *Life of John Fitch,* 348–49.

25. Thornton, "A Short Account," 16.

26. Griffin Greene was a first cousin of the Revolutionary War General Nathanael Greene.

27. Jackson, "The Philadelphia Steamboat of 1796," 204; NMAH, Land Transportation Archives. The story of Griffin Greene's steamboat came to light in 1966, when the curator of maritime history at the Smithsonian's National Museum of American History in Washington, D.C., was handed a box of thirteen documents, a gift from the Ohioan Society. From those papers, the museum was able to construct a model of Greene's steamboat and an operating model of his steam engine. The design featured a stern paddlewheel, a

horizontally mounted cylinder, and an engine "so strikingly similar to James Watt's single-acting condensing steam engine series in almost every feature . . . that mere coincidence is unlikely." Greene's name is otherwise missing from the annals of early steamboat building.

28. Virginia State Land Office grants. The records in Richmond show a grant for 300 acres on Coxe's Creek in Jefferson County (soon after renamed Nelson County) on June 1, 1782; and a grant for 1,000 acres on Coxe's Creek and 300 acres on the south bank of Simpson's Creek, dated September 1, 1782.

29. John Rowan was twenty-three years old at the time Fitch met him; he was later a judge and a U.S. senator from Kentucky. He built a house in Bardstown, called Federal Hill, which inspired Stephen Foster, a Rowan family cousin, to write the song, "My Old Kentucky Home." The house still stands, part of a Kentucky state park.

30. Boyd, *Poor John Fitch,* 290.

31. Westcott, *Life of John Fitch,* 364.

32. Ibid., 365.

33. Harris, *Papers of William Thornton,* 433–34.

34. NMAH, Fitch file, letter of Roscoe Conkling Fitch to the *Trenton Evening Times,* 18 January 1930.

Notes to Chapter 11: The French Connection

1. In 1777, Livingston was named chancellor of New York, the state's highest judicial position. In that role, he administered the presidential oath of office to George Washington in New York City in 1789. He held the post until 1801, when he was appointed minister to France.

2. Philip, *Robert Fulton,* 121.

3. Ibid., 122; Turnbull, *John Stevens,* 132.

4. Farrell, *Captain Samuel Morey,* 7.

5. Carter, *Samuel Morey,* 20–21. Morey is remembered, if rarely, for patenting a forerunner of the internal combustion engine, a machine fueled by turpentine vapors, in 1826. He installed this engine in a small boat in 1829.

6. Turnbull, *John Stevens,* 133.

7. Pursell, *Early Stationary Steam Engines,* 6.

8. NMAH, John Stevens Collection.

9. Livingston, "Invention of the Steamboat," 163.

10. Turnbull, *John Stevens,* 133.

11. NMAH, John Stevens Collection.

12. Turnbull, *John Stevens,* 162.

13. Philip, *Robert Fulton,* 23–26.

14. Ibid., 36. Stanhope had obtained English patents in 1790 for a steamboat propelled by paddles shaped like ducks' feet. (Spratt, *Birth of the Steamboat,*

54–55) In March 1793, the year Fulton began corresponding with Stanhope and just weeks after the only trial of Rumsey's *Columbian Maid* there, the earl tested his first full-scale steamboat on the Thames, which featured a steam engine of his own manufacture, since Boulton & Watt would not work with him. He was disappointed when it went no faster than three miles per hour, and he abandoned his experiments. In September 1793, Fulton claimed to have sent Stanhope a letter proposing he try a paddlewheel design.

15. Sale, *Fire of His Genius,* 58.
16. LOC, GW Papers, GW to Fulton, 14 December 1796.
17. Philip, *Robert Fulton,* 65–66.
18. Ibid., 78; Flexner, *Steamboats Come True,* 261.
19. French private ships and naval vessels had been plundering American trading ships for years, and in 1797 President Adams sent a peace delegation to France to try to resolve the problem. Instead of meeting with its members, the French Foreign Minister Talleyrand refused to see them and demanded bribes as a condition of negotiation, in the scandal that became known as the XYZ affair. Adams brought his negotiators home and canceled the U.S. treaties with France, beginning what historians call the "quasi-war" with France.
20. Sale, *Fire of His Genius,* 68–69.
21. Philip, *Robert Fulton,* 88. According to the Fulton biographer H. W. Dickinson, Fulton was a frequent visitor at Cartwright's home during the previous year in England. At the time, Cartwright was experimenting with an alcohol engine, which is what originally drew Fulton to visit him. Fulton wrote to Cartwright in 1802, when he began experimenting with steamboats, to inquire about his progress with the alcohol engine. Cartwright obtained a patent for it, but he never brought it to practical use.
22. Ibid., *Robert Fulton,* 90–91.
23. Ibid., 96–97.
24. Sale, *Fire of His Genius,* 81.
25. LOC, TJ Papers, TJ to Robert R. Livingston, 18 April 1802.
26. Philip, *Robert Fulton,* 125.
27. Sutcliffe, *Robert Fulton,* 330.
28. Dickinson, *Robert Fulton,* 144.
29. Spratt, *Birth of the Steamboat,* 62–63.
30. Dickinson, *Robert Fulton,* 149–50.
31. Ibid., 155.
32. Ibid., 157–58.
33. Ibid., 165.
34. Dabney, *Silver Sextant,* 139.
35. Sale, *Fire of His Genius,* 98.
36. Flexner, *Steamboats Come True,* 310.
37. Dickinson, *Robert Fulton,* 199.

38. Sale, *Fire of His Genius,* 110.
39. Spratt, *Birth of the Steamboat,* 62–63.
40. Turnbull, *John Stevens,* 174.
41. Gilbert, "Oliver Evans' Memoir," 155–56.
42. Ferguson, *Oliver Evans,* 36.
43. Gilbert, "Oliver Evans' Memoir," 157.
44. Turnbull, *John Stevens,* 158.
45. Gilbert, "Oliver Evans' Memoir," 160–61.
46. Turnbull, *John Stevens,* 185. In those days, the Columbia campus was located near Trinity Church in downtown Manhattan.
47. Ibid., 189. The machinery of this boat has survived and is in the collections of the Smithsonian's National Museum of American History.

Notes to Chapter 12: Steamboat Collisions

1. Sutcliffe, *Robert Fulton,* 207.
2. Sale, *Fire of His Genius,* 109.
3. Thornton, his wife, Anna Maria Brodeau, and her mother settled at first in Georgetown, in a leased house on M Street. They later moved to F Street, N.W., where their next-door neighbors were James and Dolley Madison. (Stearns and Yerkes, *William Thornton,* 27.) Thornton and Madison had lived in the same Philadelphia boardinghouse the summer of 1787, shortly after Thornton moved to America, when Madison was a delegate to the Constitutional Convention.
4. Ibid., 27–28.
5. Philip, *Robert Fulton,* 186.
6. Ibid., 189.
7. Ibid., 192.
8. Ibid., 193.
9. Ibid., 196.
10. LOC, TJ Papers, TJ to Fulton, 16 August 1807.
11. Sutcliffe, *Robert Fulton,* 234–35.
12. Dickinson, *Robert Fulton,* chapter 9.
13. Sutcliffe, *Robert Fulton,* 259.
14. Walterscheid, *To Promote the Progress,* 479–80.
15. Philip, *Robert Fulton,* 211.
16. LOC, TJ Papers, Fulton to TJ, 3 December 1807.
17. Harris, *Papers of William Thornton,* 147n.
18. Philip, *Robert Fulton,* 179.
19. NMAH, John Stevens Collection.
20. Ibid., 237–38.
21. Turnbull, *John Stevens,* 230.
22. Ibid., 254–55.
23. Philip, *Robert Fulton,* 229.

24. Ibid., 232.
25. Ibid., 225.
26. Ibid., 227; 235; Sale, *Fire of His Genius,* 136.
27. Harris and Preston, *Papers Relating to the Administration,* Thornton to John Stevens, 23 January 1809.
28. Philip, *Robert Fulton,* 234.
29. Thornton, "A Short Account."
30. Harris and Preston, *Papers Relating to the Administration,* Thornton to John Stevens, 24 November 1808.
31. Ibid., John Stevens to Thornton, 11 January 1809.
32. Ibid., Thornton to John Stevens, 23 January 1809.
33. Ibid., Thornton to John Stevens, 15 February 1809.
34. Philip, *Robert Fulton,* 237.
35. Ibid., 239.
36. LOC, Fulton Papers, Thornton to Fulton, 12 May 1809.
37. Ibid. Fulton studied Beaufoy's *Nautical Experiments,* which provided data on the resistance of various solid shapes as they move through water, to come up with his design proportions. His later steamboats had increasingly wider beams, for stability. As it turned out, given the low horsepower engines he used, the long, narrow shape he first favored had little effect on speed.
38. Turnbull, *John Stevens,* 271.
39. Ibid., 275–79.
40. Ibid., 280.
41. NMAH, John Stevens Collection, Fulton and Livingston to Stevens, 11 November 1809.
42. Ibid., 33–34, Stevens to Fulton and Livingston, 21 November 1809.
43. Ibid., 38–39.
44. Sale, *The Fire of His Genius,* 143.
45. Maryland Historical Society, Papers of Benjamin Henry Latrobe, Latrobe to Nicholas Roosevelt, 7 February 1809.
46. Hamlin, *Benjamin Henry Latrobe,* 225–30.
47. Ibid., 265.
48. Ibid., 282, n. 25. Hamlin describes an 1806 incident in which Latrobe was struck on the head by a brick falling from the Capitol's scaffolding; "No suggestion is intended that Thornton was directly involved, of course; but the uncertainties which resulted from his campaign against Latrobe may have helped to produce in some unbalanced workman an unreasoning dislike of Latrobe." Latrobe related the incident to Jefferson (LOC, TJ Papers, Benjamin Latrobe to TJ, 21 April 1806).
49. Stearns and Yerkes, *William Thornton,* 43.

NOTES TO CHAPTER 13: JOHN FITCH'S GHOST

1. Latrobe, Charles J., *Rambler,* letter 7.

2. USGS, "Whole Lotta Shakin'." Two other major earthquakes rocked the same area the following January and February.
3. Latrobe, J. H. B., "First Steamship Voyage." John Hazelhurst Boneval Latrobe was a son of Nicholas and Lydia who related the account that follows.
4. Hamlin, *Benjamin Henry Latrobe,* 594.
5. Ibid., 594–95.
6. LOC, FP, Fulton to William Thornton, 9 January 1811.
7. Philip, *Robert Fulton,* 262.
8. Ibid., 262.
9. Ibid., 262–64.
10. Sutcliffe, *Robert Fulton,* 287.
11. Woodress, *Yankee's Odyssey,* 282.
12. Sutcliffe, *Robert Fulton,* 353.
13. Philip, *Robert Fulton,* 276.
14. Ibid., 284.
15. Pursell, *Early Stationary Steam,* 35.
16. Hamlin, *Benjamin Henry Latrobe,* 596.
17. Ibid., 378.
18. Philip, *Robert Fulton,* 275.
19. Ibid., 285.
20. Ibid., 290.
21. Sale, *Fire of His Genius,* 163.
22. Philip, *Robert Fulton,* 294.
23. Turnbull, *John Stevens,* 334, 340.
24. Hamlin, *Benjamin Henry Latrobe,* 399.
25. Ibid., 597.
26. Ibid., 598–99.
27. Ibid., 599.
28. Ibid., 431.
29. Philip, *Robert Fulton,* 316.
30. Ibid., 316–17.
31. Thornton, letter to *National Intelligencer,* Washington, D.C., 7 September 1814.
32. Sale, *Fire of His Genius,* 155.
33. Ibid., 325.
34. Ibid., 155.
35. Philip, *Robert Fulton,* 335–37.
36. Ibid., 337.
37. Latrobe, J. H. B. "Lost Chapter," appendix.
38. Harris and Preston, *Papers Relating to the Administration,* Fulton to James Madison, 27 December 1814.
39. Ibid., James Madison to William Thornton, 27 December 1814.
40. Ibid., William Thornton to James Madison, 9 January 1815.

41. Ibid., Fulton to Richard Rush, 9 January 1815.
42. Latrobe, J. H. B. "Lost Chapter," John Delacy to Nicholas Roosevelt, 15 January 1815.
43. Stockton, *History of the Steam-Boat Case,* 4.
44. Philip, *Robert Fulton,* 89–92.
45. Thornton, "Short Account," 18–19.
46. Ibid., 19–20.
47. Philip, *Robert Fulton,* 343–44.
48. Ibid., 344. Another issue at the hearing was the matter of each state's rights to the Hudson River. New York claimed that its state line extended to the low-water mark on the New Jersey shore. New Jersey naturally assumed that the two states shared the river. Nothing was resolved. It would take another nineteen years for the two states to agree to set the state line down the middle of the river.
49. Stockton, *History of the Steam-Boat Case,* 31.
50. Philip, *Robert Fulton,* 345.

Notes to Epilogue

1. Hunter, "Heroic Theory," 27.
2. National Geographic Society, *Those Inventive Americans,* 18.
3. Wills, *Mr. Jefferson's University,* 91–93.; Hamlin, *Benjamin Henry Latrobe,* 470.
4. Ibid., 450, 528.
5. Turnbull, *John Stevens,* 439–40.
6. Ibid., 362.
7. This was not the world's first; British engineer Richard Trevithick put a steam locomotive to work at a Welsh ironworks in 1804.
8. Turnbull, *John Stevens,* 502.
9. Hunter, "Heroic Theory," 40.
10. High pressure in those days meant a pressure of more than 100 pounds per square inch; low-pressure engines like Watt's worked at a pressure of 15 to 20 pounds per square inch. The earth's atmospheric pressure at sea level is 14.7 pounds per square inch.
11. Ferguson, *Oliver Evans,* 45–63.
12. Stearns and Yerkes, *William Thornton,* 52, 55.
13. Examples include Moray's "Brief Account of Last Century's Inventive Steam Pirate" and Gosnell's, "First American Steamboat: James Rumsey Its Inventor, not John Fitch."
14. VHS, Rumsey Papers.
15. NMAH, Fitch folder, John S. Still, Ohio State Museum, to Alexander Uhl, Public Affairs Institute, Washington, D.C., 5 March 1954. Fitch's great-grandson was Augustus N. Whiting, son of Orrel Kilbourne Whiting, who

was a daughter of Lucy Fitch Kilbourne. Lucy had moved to Worthington, Ohio, near Columbus, in 1803. Interestingly, her husband, James Kilbourne, and his friend Thomas Worthington were agents for the Scioto Associates land company.

16. Ohio Historical Society, *Museum Echoes,* January 1955, 3.
17. NMAH, Fitch folder, letter to Alexander Uhl from John S. Still, curator of historical collections, Ohio State Museum, Columbus, 15 September 1954.
18. Hindle, *Emulation,* 77.

BIBLIOGRAPHY

Barnes, Joseph. "Remarks on Mr. John Fitch's Reply to Mr. James Rumsey's Pamphlet." Philadelphia: July 7, 1788.

Basalla, George. *The Evolution of Technology.* Cambridge: Cambridge University Press, 1988.

Bathe, Greville. *An Engineer's Miscellany.* Philadelphia: Patterson & White Company, 1938.

————. *Three Essays: A Dissertation on the Genius of Mechanical Transport in America Before 1800.* St. Augustine, FL: Published by the author, 1960.

Bedini, Silvio A. "History Corner: Henry Voigt." *Professional Surveyor Magazine* 17, no. 7 (October 1997).

Belote, Theodore. *The Scioto Speculation and the French Settlement at Gallipolis: A Study in Ohio Valley History.* New York: Burt Franklin, 1971.

Billington, David P. *The Innovators: The Engineering Pioneers Who Made America Modern.* New York: John Wiley & Sons, 1996.

Boorstin, Daniel. *The Americans: The National Experience.* New York: Random House, 1965.

Bourne, Russell. *Invention in America.* Golden, CO: Fulcrum Publishing, 1996.

Bowen, Catherine Drinker. *Miracle at Philadelphia: The Story of the Constitutional Convention.* Boston: Little, Brown & Co., 1966.

Boyd, Julian P., ed. *The Papers of Thomas Jefferson.* 28 volumes. Princeton: Princeton University Press, 1950.

Boyd, Thomas. *Poor John Fitch: Inventor of the Steamboat.* Freeport, NY: Books for Libraries Press, 1971.

Brookhiser, Richard. *Founding Father: Rediscovering George Washington.* New York: The Free Press, 1996.

Bruno, Leonard C. *Tradition of Technology: Landmarks of Western Technology in the Collections of the Library of Congress.* Washington, D.C.: Library of Congress, 1995.

Cardwell, D. S. L. *Turning Points in Western Technology: A Study of Technology, Science, and History.* Canton, MA.: Science History Publications, 1991.

Carter, George Calvin. *Samuel Morey: The Edison of His Day.* Concord, NH: Published by the author, 1945.

Conway, Moncure Daniel. *The Life of Thomas Paine.* Vol. 2. New York: G. P. Putnam's Sons, 1892.

Cowan, Ruth. *A Social History of American Technology.* New York: Oxford University Press, 1997.

Cross, Gary, and Rick Szostak. *Technology and American Society: A History.* Englewood Cliffs, NJ: Prentice-Hall, 1995.

Cruse, Englehart. "The Projector Detected, or Some Strictures on the Plan of Mr. James Rumsey's Steam Boat." Baltimore: Published by the author, 1788.

Curtis, Darwin O'Ryan. "The Rumseian Experiment." Hagerstown, MD: Published by the author, 1987.

Dabney, Betty Page. *The Silver Sextant: Four Men of the Enlightenment.* Norfolk, VA: Published by the author, 1993.

Daumas, Maurice. *A History of Technology and Invention: Progress Through the Ages. Vol. III. The Expansion of Mechanization, 1725 - 1860.* Translated by Eileen B. Hennessy. New York: Crown Publishers, 1979.

Davis, William H. *The History of Bucks County, Pennsylvania.* 2nd ed. New York: Lewis Publishing Company, 1905. Online at http://www.rootsweb.com/~usgenweb/pa/bucksp/davistoc.htm.

Davis, Wilma S., ed. *Decisions of the United States Courts Involving Copyright and Literary Property, 1789 - 1909.* Copyright Office Bulletin No. 13. Washington, D.C.: Copyright Office, Library of Congress, 1980.

Dayton, Fred Erving. *Steamboat Days.* New York: Frederick Stokes Company, 1925.

De Pauw, Linda Grant, ed. *House of Representatives Journal.* Vol. 3. Baltimore: Johns Hopkins University Press, 1977.

Dickinson, H. W. *Robert Fulton: Engineer and Artist.* Freeport, NY: Books for Libraries Press, 1971.

Dickinson, H. W. and Rhys Jenkins. *James Watt and the Steam Engine.* Oxford: Clarendon Press, 1927.

Eckhardt, George H. *Pennsylvania Clocks and Clockmakers: An Epic of Early American Science, Industry, and Craftsmanship.* New York: Devin-Adair Company, 1955.

Eco, Umberto and G. B. Zorzoli. *A Pictorial History of Inventions.* London: Weidenfeld and Nicolson, 1961.

Evans, Oliver. *The Abortion of the Young Steam Engineer's Guide.* Philadelphia: Fry and Kammerer, 1805.

———. *The Young Mill-wright and Miller's Guide.* New York, Arno Press, 1972.

———. *The Young Steam Engineer's Guide.* Philadelphia: H. C. Carey and I. Lea, Chestnut Street, n.d.

Farrell, Gabriel Jr. *Captain Samuel Morey, Who Built a Steamboat Fourteen Years Before Fulton.* Manchester, NH: Standard Book Company, 1915.

Federico, P. J. "Operation of the Patent Act of 1790." *Journal of the Patent Office Society* 18, no. 4 (April 1936): 237–251.

Ferguson, Eugene S. *Oliver Evans: Inventive Genius of the American Industrial Revolution.* Greenville, DE: The Hagley Museum, 1980.

Fitch, John. "The Original Steamboat Supported; or, a Reply to Mr. James Rumsey's Pamphlet Shewing the True Priority of John Fitch and the False Datings, Etc. of James Rumsey." Philadelphia: Published by the author, May 1788.

Fitch, Roscoe Conkling. *History of the Fitch Family.* Haverhill, MA: Record Publishing Company, 1930.

Fitzpatrick, John C., ed. *The Writings of George Washington from the Original Manuscript Sources, 1745 - 1799.* 39 vols. Washington, D.C.: Government Printing Office, 1931–44. Online at http://memory.loc.gov/ammem/gwhtml/gwhome.html.

Flexner, James Thomas. *Steamboats Come True: American Inventors in Action.* New York: Fordham University Press, 1992.

Ford, Edward. *David Rittenhouse: Astronomer-Patriot.* Philadelphia: University of Pennsylvania Press, 1946.

Freeman, Douglas Southall. *George Washington: A Biography.* Vol. 6: *Patriot and President.* New York: Charles Scribner's Sons, 1968.

Friedenberg, Daniel M. *Life, Liberty and the Pursuit of Land: The Plunder of Early America.* Buffalo, NY: Prometheus Books, 1992.

"Gallipolis." Book review, *Transatlantic 1/2000.* E-journal. American Studies Journal, Association Française d' Etudes Americaines. Paris, France. Online at http://etudes.americaines.free.fr/transatlantica.

Gilbert, Arlan K. "Oliver Evans' Memoir 'On the Origin of Steam Boats and Steam Wagons.'" *Delaware History* 7, no. 2 (September 1956): 142–167.

Gosnell, Harpur Allen. "The First American Steamboat: James Rumsey Its Inventor, not John Fitch." *Virginia Magazine of History and Biography* 40 (1932): 124–132.

Green, James. "Report of the Librarian." In *The Annual Report of the Library Company of Philadelphia for the Year 1990.* Philadelphia: Library Company of Philadelphia, 1991.

Hamlin, Talbot. *Benjamin Henry Latrobe.* New York: Oxford University Press, 1955.

Hardenberh, Horst O. *Samuel Morey and His Atmospheric Engine.* Warrendale, PA: Society of Automotive Engineers, 1992.

Harris, C. M. "The Improbable Success of John Fitch." *Invention & Technology Magazine* (Winter 1989): 24–31.

Harris, C. M. and Daniel Preston. *Papers Relating to the Administration of the U.S Patent Office During the Superintendency of William Thornton, 1802 - 1828.* Microform. Washington, D.C.: National Archives and Records Administration, 1987.

Harris, C. M. ed. *Papers of William Thornton.* Vol. 1: 1781–1802. Charlottesville: University Press of Virginia, 1995.

Hindle, Brooke. "James Rumsey and the Rise of Steamboating in the United States." *West Virginia History* 48 (1989): 33–42.

———. *David Rittenhouse.* Princeton: Princeton University Press, 1964.

———. *Emulation and Invention.* New York: W. W. Norton & Company, 1981.

———. *Pursuit of Science in Revolutionary America, 1735 - 1789.* Published for the Institute of Early American History and Culture, Williamsburg, VA. Chapel Hill: University of North Carolina Press, 1956.

—. *Technology in Early America: Needs and Opportunities for Study.* Published for the Institute of Early American History and Culture, Williamsburg, VA. Chapel Hill: University of North Carolina Press, 1966.

Hindle, Brooke, and Stephen Lubar. *Engines of Change.* Washington, D.C.: Smithsonian Institution Press, 1986.

Hodgson, Alice Doan. *Samuel Morey: Inventor Extraordinary.* Orford, NH: Historical Fact Publications, 1961.

Hulbert, Archer Butler. *Washington and the West: Being George Washington's Diary of September 1784.* New York: The Century Company, 1905.

Hunter, Louis. "The Heroic Theory of Invention." In *Technology and Social Change in America.* Ed. Edwin T. Layton, Jr., 25–46. New York: Harper & Row, 1973.

Hunter, Marshall T. "James Rumsey." *Virginia Cavalcade,* Autumn 1964.

Inland Waterways Association of Ireland. "The Royal Canal." 4th ed. Waterways Service of the Department of Arts, Culture, and the Gaeltacht, Dublin, Ireland: 1997. Online at http://www.iwai.ie/maps/royal/contents.html.

Jackson, Donald, and Dorothy Twohig, eds. *The Diaries of George Washington,* 6 vols. Charlottesville: University Press of Virginia, 1976–79. George Washington, Diary, September 1–October 4, 1784. Online at http://memory.loc.gov/ammem/gwhtml/gwhome.html.

Jackson, Melvin H. "The Philadelphia Steamboat of 1796." *The American Neptune* 50, no. 3 (Summer 1990): 201–210.

James, J. G. "Thomas Paine's Iron Bridge Work, 1785–1803." *Transactions* 59 (1987–88). Newcomen Society for the Study of the History of Engineering and Technology. London: 1990.

Jordan, Francis Jr. *The Life of William Henry of Lancaster, Pennsylvania.* Lancaster: New Era Printing Company, 1910.

Journals of the Continental Congress, 1774–1789. Ed. Worthington C. Ford et al. Washington, D.C., 1904–37, 19:137. Online at http://memory.loc.gov/ammem/amlaw/lwjc.html.

Journal of the Patent Office Society. "Proceedings in Congress During the Years 1789 and 1790, Relating to the First Patent and Copyright Laws. 22 (April 1940): 243–278.

Kasson, John F. *Civilizing the Machine: Technology and Republican Values in America, 1776–1900.* New York: Hill and Wang, 1999.

Keane, John. *Tom Paine: A Political Life.* New York: Little, Brown & Co., 1995.

Kemp, Emory. "James Rumsey and His Role in the Improvements Movement." *West Virginia History* 48 (1989):1–5.

Kenamond, A. D. "James Rumsey and His Steamboat." *Magazine of the Jefferson County Historical Society* 3 (December 1937): 4–11.

Kranzberg, Melvin, and Carroll W. Pursell, Jr. *Technology in Western Civilization.* Vol. 1. New York: Oxford University Press, 1967.

Laidley, W. S. "James Rumsey: The Inventor of the Steamboat." *West Virginia Historical Magazine* 3, no. 1 (January 1903): 185–190.

Latrobe, Benjamin Henry. "First Report of Benjamin Henry Latrobe, . . . in answer to the question of the Society of Rotterdam, whether any, and what, improvements have been made in the construction of steam engines in America?" *Transactions of the American Philosophical Society* 6 (1809): 89–98.

Latrobe, Charles J. *The Rambler in North America, 1822 - 1833.* London: 1836. Baltimore, Maryland Historical Society, Papers of Benjamin Henry Latrobe. Extract online at http://www.myoutbox.net/nr1836.htm.

Latrobe, J. H. B. "A Lost Chapter in the History of the Steamboat." *Maryland Historical Society Fund Publication* 5 (1871). Online at http://www.myoutbox.net/nr1871a.htm.

———. "The First Steamboat Voyage on the Western Waters." *Maryland Historical Society Fund Publication,* October 1871. Online at http://www.myoutbox.net/nr1871b.htm.

Layton, Edwin T., Jr. "James Rumsey: Pioneer Technologist." *West Virginia History* 48 (1989): 6–32.

———. "The Most Original." *Invention and Technology* (Spring 1987): 50–56.

Livingston, Robert R. "The Invention of the Steamboat: An Historical Account of the Application of Steam for the Propelling of Boats." Letter from Chancellor Livingston to the editors of the *American Medical and Philosophical Register,* published in that journal in January 1812. Reprinted in Old South Leaflets No. 108. Boston: Directors of the Old South Work, 1902.

Lloyd, James T. *Lloyd's Steamboat Directory, and Disasters on the Western Waters.* Cincinnati: J. T. Lloyd & Company, 1856.

Marestier, Jean Baptiste. *Memoir on Steamboats of the United States of America.* Translated by Sidney Withington. Mystic, CT: Marine Historical Association, Inc., 1957.

McFarlane, Robert. *History of Propellers and Steam Navigation: With Biographical Sketches of the Early Inventors.* New York: G. P. Putnam, 1851.

McGaw, Judith A. *Early American Technology: Making and Doing Things from the Colonial Era to 1850.* Published for the Institute of Early American History and Culture, Williamsburg, VA. Chapel Hill: University of North Carolina Press, 1994.

Moray, John. "A Brief Account of Last Century's Inventive Steam Pirate." Berkeley Springs, WV: 1910.

Morrison, John. H. *History of American Steam Navigation.* New York: Stephen Daye Press, 1958.

Mowry, William A. *Who Invented the American Steamboat?* Bristol: New Hampshire Antiquarian Society, 1874.

National Geographic Society. *Those Inventive Americans.* Washington, D.C.: National Geographic Society, 1971.

New York Assembly Papers, Miscellaneous, Vol. 3.1. Petition of the Rumseian Society, Philadelphia, to the Speaker of the Assembly New York. September 23, 1788.

Oliver, John W. *History of American Technology.* New York: Ronald Press Company, 1956.

Parsons, Mira Clarke. "John Fitch, Inventor of Steamboats." *Ohio Archaeological and Historical Society Quarterly* 8 (1900): 397–408.

Pennsylvania State Archives. Records of Nicholson Lands, Articles of Impeachment, 1792, series 17.219. Records of the Land Office, RG-17.

Perkins, Sid. "When Horses Really Walked on Water." *The Chronicle of the Horse.* (May 20, 1999): 90–92.

Philip, Cynthia Owen. *Robert Fulton: A Biography.* New York: Franklin Watts, 1985.

Prager, Frank D. "An Early Steamboat Plan of John Fitch." *The Pennsylvania Magazine of History and Biography* 79 (January 1955): 63–81.

———. "Historic Background and Foundation of American Patent Law." *American Journal of Legal History* 5 (1961): 309–325.

———. "Proposal for the Patent Act of 1790." *Journal of the Patent Office Society* 36 (March 1954): 157–167.

———. "The Steam Boat Interference." *Journal of the Patent Office Society* 40 (September 1958): 611–643.

———. "The Steamboat Pioneers Before the Founding Fathers." *Journal of the Patent Office Society* 37 (July 1955): 486–522.

Prager, Frank D., ed. *The Autobiography of John Fitch.* Philadelphia: The American Philosophical Society, 1976.

Pursell, Carroll. *Early Stationary Steam Engines in America.* Washington, D.C.: Smithsonian University Press, 1969.

———. *The Machine in America: A Social History of Technology.* Baltimore: Johns Hopkins University Press, 1995.

———. *Readings in Technology and American Life.* New York: Oxford University Press, 1969.

Pursell, Carroll, ed. *Technology in America: A History of Individuals and Ideas.* Cambridge, MA: MIT Press, 1981.

Rasmussen, William M. S., and Robert S. Tilton. *George Washington: The Man Behind the Myths.* Charlottesville: University Press of Virginia, 1999.

Read, David. *Nathan Read.* New York: Hurd and Houghton, 1870.

Rollins, Richard M., ed. *The Autobiographies of Noah Webster: From the Letters and Essays, Memoir, and Diary.* Columbia: University of South Carolina Press, 1989.

Rumsey, James. "Explanation of a Steam Engine, and the Method of Applying It to Propel a Boat" and "The Explanations and Annexed Plates of the Following Improvements to Mechanics." Philadelphia: Published by the author, 1788.

———. "A Short Treatise on the Application of Steam, Whereby is Clearly Shewn from Actual Experiments. . . ." Philadelphia: Published by the author, January 1788.

Sale, Kirkpatrick. *The Fire of His Genius: Robert Fulton and the American Dream.* New York: The Free Press, 2001.

Scharf, J. Thomas, and Thompson Westcott. *History of Philadelphia, 1609–1884.* 3 vols. Philadelphia: L. H. Everts & Co., 1884.

Singer, Charles, E. J. Holmyard, A. R. Hall, and Trevor I. Williams. *A History of Technology.* Vol. IV. *The Industrial Revolution.* Oxford: Clarendon Press, 1958.

Smith, Merritt Roe, and Gregory Clancey. *Major Problems in the History of American Technology.* Boston: Houghton Mifflin, 1998.

Smith, Paul H., et al., eds. *Letters of Delegates to Congress, 1774–1789.* 25 volumes, Washington, D.C.: Library of Congress, 1976–2000. Online at http://memory.loc.gov/ammem/amlaw/lwdg.html.

Smith, P. C. F. *The Empress of China.* Philadelphia: Maritime Museum, 1984.

Spratt, H. Philip. *The Birth of the Steamboat.* London: Charles Griffin & Company, 1958.

Stearns, Elinor, and David N. Yerkes. *William Thornton: A Renaissance Man in the Federal City.* Washington, D.C.: American Institute of Architects, 1976.

Stockton, Lucius Horatio. *A History of the Steam-Boat Case, lately discussed by counsel before the legislature of New Jersey: comprised in a letter to a gentlemen in Washington.* Trenton, NJ: 1815. Microform. Early American Imprints, second series, no. 36022.

Sutcliffe, Ann Crary. *Robert Fulton and the Clermont.* New York: The Century Company, 1909.

Thornton, William. "Short Account of the Origin of Steam Boats." Washington, D.C.: Published by the author, 1814.

Todd, Charles Burr. *Life and Letters of Joel Barlow: Poet, Statesman, Philosopher.* New York: De Capo Press, 1970.

Trevor, Williams. *The History of Invention.* New York: Facts on File, Inc., 1987.

Turnbull, Archibald. *John Stevens: An American Record.* New York: The Century Company, 1928.

Turner, Ella May. *James Rumsey: Pioneer in Steam Navigation.* Scottsdale, PA: Mennonite Publishing House, 1930.

Uglow, Jenny. *The Lunar Men: Five Friends Whose Curiosity Changed the World.* New York: Farrar Straus and Giroux, 2002.

Unger, Harlow G. *Noah Webster: The Life and Times of an American Patriot.* New York: John Wiley & Sons, 1998.

U.S. Geological Survey. "The Mississippi Valley: 'Whole Lotta Shakin' Goin' On.'" Factsheet 168–95–1995, New Madrid. Online at http://quake.wr.usgs.gov/prepare/factsheets/NewMadrid.

Walterscheid, Edward C. "Defining the Patent and Copyright Term: Term Limits and the Intellectual Property Right Clause." *Journal of Intellectual Property Law* 7 (Spring 2000): 347–356.

———. "The Early Evolution of United States Patent Law." *Journal of the Patent and Trademark Society* (October 1996): 672–685.

———. *The Nature of the Intellectual Property Clause: A Study in Historical Perspective.* Buffalo, NY: W. S. Hein & Co., 2002.

———. *To Promote the Progress of Useful Arts: American Patent Law and Administration.* Littleton, CO: F. B. Rothman, 1998.

Watson, John F. *Watson's Annals of Philadelphia and Pennsylvania,* 1857, Vol. 1, chapter 90: Persons and Characters, part 3: John Fitch. Online at http://ftp.rootsweb.com/pub/usgenweb/pa/philadelphia/areahistory/watson0119.txt.

————. *Watson's Annals of Philadelphia and Pennsylvania,* 1857, Vol. 2, chapter 30: Steamboats, "John Fitch." Online at http://ftp.rootsweb.com/pub/usgenweb/pa/ philadelphia/areahistory/watson0215.txt.

Westcott, Thompson. *The Life of John Fitch, the Inventor of the Steamboat.* Philadelphia: J. B. Lippincott & Company, 1878.

White, Lynn, Jr. *Medieval Technology and Social Change.* Oxford: Clarendon Press, 1962.

Whittlesey, Charles. "Justice to the Memory of John Fitch." *Western Literary Journal and Monthly Review* (February 1845).

William & Mary Quarterly Historical Magazine. "Letters of James Rumsey, Inventor of the Steamboat." Vol. 24, issue 3 (January 1916): 154–174.

William & Mary Quarterly Historical Magazine. "Letters of James Rumsey, Inventor of the Steamboat." Vol. 24, issue 4 (April 1916): 239–251.

William & Mary Quarterly Historical Magazine. "Letters of James Rumsey, Inventor of the Steamboat." Vol. 25, issue 1 (July 1916): 21–34.

Williams, Trevor I. *The History of Invention.* New York: Facts on File, 1987.

Wills, Garry. *Mr. Jefferson's University.* Washington, D.C.: National Geographic Society, 2002.

Woodcroft, Bennet. *A Sketch of the Origin and Progress of Steam Navigation from Authentic Documents.* London: Taylor, Walton, and Maberly, 1848.

Woodress, James. *A Yankee's Odyssey: The Life of Joel Barlow.* Philadelphia: J. B. Lippincott Company, 1958.

York, Neil Longley. *Mechanical Metamorphosis: Technological Change in Revolutionary America.* Westport, CT: Greenwood Press, 1985.

Index